The Biblical Response to Aborted Fetal Medicine

The Lies That Claim Its Necessity, & The Truth About Its Consequences

The Biblical Response to Aborted Fetal Medicine
The Lies That Claim Its Necessity, & The Truth About Its Consequences

This content is not intended to be a substitute for professional medical advice, diagnosis, or treatment. Always seek the advice of your physician or other qualified health provider with any questions you may have regarding a medical condition.

The original emphasis in excerpts and quotations has been preserved with bold and italicized fonts; the author's emphasis is underlined. If an entire quote was emboldened or italicized it has been changed to regular font in this publication.

Many websites have been quoted; If they change in the future, it should be noted that these references are accurate at the time of publishing this book.

All Scripture quotations are taken from the King James Version.

ALPHA AND OMEGA MEDIA, LLC
www.AlphaAndOmega.com

All profits from this book will be donated to missionaries faithfully sharing the gospel throughout the world.

ISBN 978-1-7378104-9-0

Table of Contents

Malachi 2:17 **Ye have wearied the LORD with your words. Yet ye say, Wherein have we wearied him? When ye say, Every one that doeth evil is good in the sight of the LORD, and he delighteth in them; or, Where is the God of judgment?**

Isaiah 33:15 **He that walketh righteously, and speaketh uprightly; he that despiseth the gain of oppressions, that shaketh his hands from holding of bribes, that stoppeth his ears from hearing of blood, and shutteth his eyes from seeing evil;**

For those who do not care about being right, but rather love what is right.

Part I

Isaiah 34:16 **Seek ye out of the book of the LORD, and read:**

Psalm 143:8 **Cause me to hear thy lovingkindness in the morning; for in thee do I trust: cause me to know the way wherein I should walk; for I lift up my soul unto thee.**

"I am pursuing Truth, and while I can keep my eye fixed upon my guide, I am indifferent whither I am led, provided she is my leader."

—Dr. Benjamin Rush

The "Birth" of Israel

Ezekiel 16:4 **And as for thy nativity, in the day thou wast born thy
navel was not cut, neither wast thou washed in water to supple
thee; thou wast not salted at all, nor swaddled at all.** 5 **None eye
pitied thee, to do any of these unto thee, to have compassion upon
thee; but thou wast cast out in the open field, to the lothing of thy
person, in the day that thou wast born.** 6 **And when I passed by
thee, and saw thee polluted in thine own blood, I said unto thee
when thou wast in thy blood, Live; yea, I said unto thee when thou
wast in thy blood, Live.** ...

8 **Now when I passed by thee, and looked upon thee, behold, thy
time was the time of love; and I spread my skirt over thee, and
covered thy nakedness: yea, I sware unto thee, and entered into a
covenant with thee, saith the Lord GOD, and thou becamest mine.**

Introduction

God's Word gives us a beautiful picture of the "adoption" of His people, the Israelites (Rom 9:4).

This world treats helpless children much differently.

It is legal to kill unwanted children, and man has declared that experimenting on their remains could lead to advances in scientific and medical knowledge which save lives. To the Biblical Christian I ask: is it reasonable to expect murder to save lives?

Matthew 7:18 **A good tree cannot bring forth evil fruit, neither can a corrupt tree bring forth good fruit.**

1 Thessalonians 5:20 **Despise not prophesyings. 21 Prove all things; hold fast that which is good. 22 Abstain from all appearance of evil.**

Scientists and medical professionals have performed atrocities by experimenting on unwanted children in the name of the "greater good." So many people are unaware they have personally used the products of these experiments by receiving vaccines made from aborted fetal cell lines. While not all vaccines are made this way, several routine childhood immunizations are made from these cell lines. When I discovered this horrifying fact, I felt as though I couldn't warn other Christians because vaccines *are not to be questioned* in today's cultural climate. Something is very wrong when a practice involving abortion cannot be questioned, but God's words *"thou shalt not kill"* can. This ought not to be, especially among Christians who are to live by *"every word of God"* (Luk 4:4).

It is not my goal to unnecessarily sow division among Christians in a culturally controversial area. Abortion is not an issue in which we have freedom; it is not a grey area. If you are a Christian, I want to warn you about a common practice most do not know is associated with abortion. I want to inform you about its consequences, and show you a better way. We will view all of this information through the lens of Scripture.

Psalm 19:9b ...**the judgments of the LORD are true and righteous altogether.** ... 11 **Moreover by them is thy servant warned: and in keeping of them there is great reward.**

Most doctors do not give true informed consent regarding vaccination. Inconvenient facts, such as aborted fetal cell lines, are not mentioned. We must ask ourselves: if this appalling association between vaccines and abortion has been hidden, what else has been concealed? What else are parents not told concerning a vaccination their child receives from a doctor? My daughter's pediatrician did not tell me any of this information; I had to go to other sources to properly understand this issue. Hundreds of studies contradicted what my child's doctor stated, and there were government classifications and studies published by the American Academy of Pediatrics, of which this physician was not aware. She was not a negligent or uncaring doctor; unfortunately, this lack of knowledge is displayed by many pediatricians. They wear blinders concerning this matter, and following their groupthink advice has brought harm to many children. The following is a compilation of the most striking information found when I researched what was best for my own child. I believe you too should know this information before you follow our government's recommendation to receive these vaccines.

Through my study, I held fast to God's law and trusted His Word above everything and it has led to life and peace. I could not concede that having even the faintest association with abortion was the right thing for my family. I would not accept that using an aborted fetal cell line is what God would have us do. Throughout my search, I compared everything to the Bible, and have compiled this information for those who hunger and thirst for something better than what this world offers. Choosing to use these vaccines is not as removed from the act of abortion as one might think; there is actually still "matter," namely DNA and protein, from the original abortions *in the final vaccine products*. This first-hand contact with aborted fetal DNA has negative effects on the living recipient.

Proverbs 3:5 **Trust in the LORD with all thine heart; and lean not unto thine own understanding.** 6 **In all thy ways acknowledge him, and he shall direct thy paths.** 7 **Be not wise in thine own eyes: fear the LORD, and depart from evil.** 8 **It shall be health to thy navel, and marrow to thy bones.**

Departing *from evil* brings health; while sin *withholds good* things from us (Jer 5:25). According to Deuteronomy 32:29, God is the only One who heals. We cannot expect to bring health or healing by violating His law. Vaccines made from aborted fetal cell lines are not only inferior to natural immunity, they are outright harmful to your children's health.

Psalm 119:165 **Great peace have they which love thy law: and nothing shall offend them.**

Please know it is not my goal to offend but to inform; to declare the righteousness of God's law, and provoke you to follow it. I'm so thankful someone made a bold statement that contradicted what *I thought I knew* about vaccines and prompted me to research this issue before my child received any vaccines made from aborted fetal cell lines. Many false claims about the benefits of these vaccines have been stated for decades; I am going to quote those who boldly tell the truth. According to God's Word, it is folly and shame to answer a matter before we hear it (Pro 18:13); please hear before you reach any conclusions.

I am not a medical professional, writer, researcher, or scientist. I'm just a parent who began researching and was horrified by what I found. I come not with excellent speech, nor with my own wisdom; I do not claim to be anyone important or have any special knowledge. I am only determined to know that God's Word is true, and to trust it implicitly. *"Though I be nothing,"* (2Co 12:11) the following contents are not relevant because of who is presenting them. These things are important because God's words are completely true. Please look past my inexperienced writing to the truth which will be plainly spoken; the Bible's emphasis on the importance of truth cannot be overstated (Psa 146:5-6; Jhn 17:17).

I am not capable of wielding words in the same way I enjoy them—deeply stirring and feeding the soul or awakening one's consciousness in the way that only perfectly chosen words can. Man's ultimate source of these words is the Scriptures, to which I will reference extensively. I will also quote those who are capable of enlightening with their wise words, and I do believe I am capable of cutting to the core of this issue with Biblical logic. I hope this combination of truth, wisdom, and logic fully illuminates this issue and leads you to a profound trust in God's Word.

It takes some time to research before one discovers the connection between aborted fetal cell lines and vaccines. Usually, most books and documentaries only mention this fact in passing, or not at all. As a Christian, this connection is what I would want to know about before anything else regarding vaccines, which is why it is the main subject of this book. Many documentaries and books are available that mention much of the scientific vaccine information which will be reviewed, but to my knowledge, none of them examine these facts from a Biblical perspective.

Many Christians, myself included, have participated in the practice of vaccination ignorantly. God gave us His laws and instruction for a reason (*"Prove all things,"* 1Th 5:21); He wants to keep us from lies that will harm us. Christians should not fear their children will become ill if they obey God's commands by refusing to have anything to do with abortion. It's ludicrous for a Christian to be unable to warn other Christians about this matter because our culture esteems vaccines to be above reproach; at the same time, this is not an easy conversation to have. That is why this information has been gathered:

to expose the deceitful, enticing wisdom of this world (1Co 2:4; Jas 3:15), and instead show you the true wisdom of God's Word (Jas 3:17).

Because I do not have any formal education on this subject, much of the content that follows is excerpted from studies, books, and quotes from very informed and educated experts; and of course, as has already been demonstrated, from God's revealed Word—the Bible. What statements I do give are based on the Scriptures and years of research. My only agenda is to believe and speak truth. I believe God's Word is our sole authority when it comes to not only faith and practice, but in all issues of this life (2Pe 1:3).

Chapter 1 will lay the Biblical foundation that murdering children is unacceptable to God; there is no reconciliation between God's Word and abortion, no matter the reasoning. Chapter 2 will show the truth behind the deceiving statements made about aborted fetal cell lines. Chapter 3 will then look at what God said would keep His people from disease, and secular knowledge which demonstrates this truth. Chapter 4 contains numerous studies showing the many negative health consequences associated with using vaccines. Chapter 5 will examine a very interesting pilot study on vaccinated versus unvaccinated children. In Chapters 6-7 there will be additional studies and information to encourage you to "Wait and Pray," and "Seek to the Lord" regarding this issue which greatly affects our children's health. Chapter 8-9 will look at the Scriptures to see the warnings God gives in regard to the wickedness of this world and the many ways in which we will be rewarded for trusting Him. Chapters 10-11 will demonstrate some of the spiritual wickedness in high places (Eph 6:12) while comparing what the Bible says about an idol called Nehushtan and the days of Noah, to what is happening in the world today. Chapter 12 will sum up the vaccine issue while we "Speak the Truth in Love," and Chapters 13-14 will finish with other truths from God's Word. The ultimate goal is to increase your trust in the Bible; everything written is a means to that end.

This is an incredibly vast and hotly debated topic, with both sides claiming to be more scientific than the other. That is why this book is starting from a *Biblical* perspective. It only takes half of one misquoted verse removed from its context to convince parents to do what the world says and use these vaccines ("God uses all things for good"); it takes much more to show them they should not participate in such a practice. What follows in chapters 1-7 is the bare minimum one should know on this subject; chapters 8-14 contain additional relevant information for those with a Biblical worldview.

All of this research has been summarized for those who were in the same place I was when this information was needed, and I encourage you to put yourself in this mindset while you read: you have just been given your first child, you have doctors telling you that you must participate in something God says "*thou shalt not*" do or your child is likely to die; you have a newborn to care for and a world full of conflicting information about vaccines to try to navigate. This content is meant to be a life preserver to keep your head above water and a lighthouse to guide you to safety. It is a guided tour, if you will, of the most relevant information that could be found on this issue from the Bible, combined with many published studies and books. This book was made as a resource for those who believe every Word of God and live based on the truths found therein. There is also plenty of science for those who don't understand how sure the Word of God is, and for those of us who enjoy the scientific echo of the power and truth of God's Word. Hopefully, this juxtaposition of what God said and what we are seeing with our own eyes will inspire an abiding trust in the Bible.

2 Timothy 3:16 **<u>All scripture</u> is given by inspiration of God, and <u>is profitable</u> for doctrine, for reproof, for correction, for instruction in righteousness:** 17 **That the man of God may be perfect, throughly furnished unto all good works.**

I will list many Old Testament verses which show how God instructed and frequently chastised the Jews. In context, these Scriptures are from a different dispensation and were directed specifically to Israel, God's adopted child and the apple of His eye (Zec 2:8). However, all things were written for our learning (Rom 15:4), and since God does not change (Mal 3:6), we can learn about what is right from the only One who is righteous (1Jo 2:1; Rev 15:4). The New

Testament also states that the things which happened to Israel were for *our* ensamples, and they were written for *our* admonition—"*upon whom the ends of the world are come*" (1Co 10:11).

As healing only comes from God, Jesus' healing of the multitudes was a clear sign of His deity. Luke 6:19 tells us: "*And the whole multitude sought to touch him: for there went virtue out of him, and healed them all.*" Jesus healed with His virtue; egregious sins will never bring health or healing. While man says evil is a means to accomplish good (Rom 3:8), God's Word warns that mankind goes against His laws to our "*own hurt*" (Jer 25:7). God's law is "*the truth*" (Psa 119:142; 151), and anything that would convince us to ignore God's righteous command "*thou shalt not kill*" is the opposite of truth. As Bible-believing Christians, why would we trust those who use the abstract, theoretical "for the greater good" ideology to justify the evil they are perpetrating with their own hands?

Many Christians and parents are wary of the newly developed Covid-19 vaccines. Most do not realize the questionable practices regarding these vaccines have already occurred with other vaccinations. Nearly everything parents dislike about the Covid-19 shots has taken place with the same vaccines they enthusiastically let their children receive. History is truly repeating itself. Please follow God's commands from 1 Thessalonians 5:20-22. If we don't know these truths, we will continue to harm ourselves and our children in our ignorance. The Bible tells us a lack of knowledge destroys (Hos 4:6), and those who are wise are known for their willingness to listen (Pro 1:5). Furthermore, the Scriptures declare it is woeful to be wise in our own eyes (Isa 5:21).

Proverbs 3:7 **Be not wise in thine own eyes: fear the LORD, and depart from evil.**

Hold fast to God's Word, hold fast that which is good, and *let go of what is not.*

"I desire to treat this important topic, not in a spirit of conscientiousness, but with a sincere desire to get at the truth even though in so doing some unwelcome facts are disclosed and some cherished ideals are dispelled. My aim is to consider facts as I find them and not as I wish they were."[1]

"To make a responsible decision in the best interests of the child, one must be willing to question a lifetime of conditioning."[2]

"No man, conscious of his moral integrity, intellectual ability and studied efforts to benefit humanity, will ever apologize for speech or book; hence I make no apology for publishing the following pages. … The general public are not aware; the householders of the land have not given this subject that attention which, as parents and guardians of little children, it is their solemn duty to do. I send forth this book to open their eyes, to rouse their conscience and to discover to them a cruel and insidious enemy where they have been cajoled into the belief they have a friend. … Error is mortal and must die; truth alone is immortal, eternal."[3]

Chapter 1

Thou Shalt Not Kill

Proverbs 13:14 **The law of the wise is a fountain of life, to depart from the snares of death.**

Psalm 119:136 **Rivers of waters run down mine eyes, because they keep not thy law.**

"Human embryos of two and one-half to five months gestation were obtained from the gynaecological department of the Toronto General Hospital. ... No macerated specimens were used and in many of the embryos the heart was still beating at the time of receipt in the virus laboratory." Drs. Thicke, Duncan, Rhodes[1]

"Embryos of between 12-18 weeks gestation have been utilized. Rarely tissues were obtained from stillborn fetuses, or from premature infants at autopsy... In the experiments...three sorts of embryonic materials were used: elements of skin, connective tissue, muscle; intestinal tissue; brain tissue. Whenever possible the embryo was removed from the amniotic sac.., transferred to a sterile towel and kept at 5 C until dissected." Dr. Enders, Weller, et al Nobel Prize for polio research[2]

"And one issue, and it tends to be a little contentious, is 'Are fetal cells used to make vaccines?' The answer's, yes. There were two elective abortions that were performed in the early 1960s, one in Sweden, one in England. Those cells that were obtained from those elective abortions have been used to make several vaccines. They've been used to make the hepatitis A vaccine, the chickenpox vaccine, the rubella, or German measles vaccine, one of the rabies vaccines, and the polio vaccine. So, it's the same cells that were obtained in the early 1960s, so no new therapeutic or elective abortions have been performed since then for the purpose of making vaccines. But the answer to the question, 'Is there a residual or very small, small, small quantities of trace DNA from those original abortions in the vaccines that I just mentioned? The answer is yes.'"[3]

Paul Offit from The Children's Hospital of Philadelphia

Genesis 4:10 **And he said, What hast thou done? the voice of thy brother's blood crieth unto me from the ground.**

It's an undeniable fact that some vaccines are made by growing viruses in human cells—electively aborted human fetal cells. Obviously, the parents elected this, as the child had no choice, even though freedom of choice is supposedly what abortion is about. Also, there is an even more disturbing fact that trace amounts of aborted fetal DNA remain and are injected along with everything else in a vaccine.

Dr. Paul Offit of the Children's Hospital of Philadelphia, states, "no new therapeutic or elective abortions have been performed since then for the purpose of making vaccines." While men who profit from vaccines might be willing to make technically true statements to mislead the masses, I'm not. The truth is, many more children have been aborted and experimented on for the research and development of these cell lines. It is possible the original research and subsequent experimentation were not performed for the express purpose of making vaccines, but great evil has been done in the name of vaccine research, which will be examined in greater detail. After all, people who support the murder of children might not have moral qualms about being dishonest. "For the greater good" allows for *much* evil.

If it were true that "only two" abortions were performed, what would be the Christian's Biblical response to this? Is the murder of two innocent children, intending to save many more lives, acceptable to God? Let's look at several verses which show how precious children are to God, and how much God hates the murder of children.

Zechariah 8:3-5 tells us Zion will be *"full of boys and girls"* playing in the streets. Psalm 127:3 says children are an *"heritage of the LORD,"* and Proverbs 17:6 calls grandchildren the *"crown"* of old men. When infants (Luk 18:15) and young children (Mar 10:13-14) were brought to Christ, the disciples rebuked those that brought them, and Jesus was *"much displeased."* The Lord wanted the children to come to Him; He picked them up and blessed them. Jesus also said, *"But whoso shall offend one of these little ones which believe in me, it were better for him that a millstone were hanged about his neck, and that he were drowned in the depth of the sea"* (Mat 18:6).

In 2 Kings 24:3-4, the Bible states that Judah was punished for its wickedness, specifically for the innocent blood that was spilled, *"which the LORD would not pardon."* Jeremiah 7:31 tells us the Israelites sacrificed their children by fire, which God had not commanded, nor had it even entered into His heart. Jeremiah 32:35 speaks of children offered to Molech, which God again says He never commanded, nor did it come to His mind. Ezekiel 11:6 says the streets were filled with the slain, and later in verse 18, this is called an abomination. Proverbs 6:16-17 tells us seven things the LORD hates and are an abomination unto Him, including hands that shed *"innocent blood."* Lastly, the only verse we really need on this matter: Deuteronomy 5:17, *"Thou shalt not kill."* Christians are to have no part with works of darkness, but instead reprove them (Eph 5:11).

Exodus 23:7 **Keep thee far from a false matter; and the innocent and righteous slay thou not: for I will not justify the wicked.**

God does not pardon the murder of innocents, regardless of the reason. Just as pagans believed human sacrifice was required to appease their gods and preserve their people, many believe these abortions are necessary to keep diseases from ravaging mankind—a modern spin of an ancient lie. Today, we sacrifice children to the gods of "Science" (falsely so called, 1Ti 6:20), "Mother Nature," and "Humanism." Killing children "for the greater good" is a pagan ideology (2Ki 17:17). God specifically calls this practice an abomination (2Ki 16:3). Isaiah 33:15 tells us that a righteous man *"stoppeth his ears"* from the hearing of blood. If someone claims that taking one child's life could somehow save billions of lives, a righteous man will not listen to such a claim.

Christ said in Matthew 18:14: *"...it is not the will of your Father which is in heaven, that one of these little ones should perish."* Man claims two abortions were required to make vaccines to supposedly save mankind. Men say "just two," while God says not even one: and unlike man, God actually means what He says (Tit 1:2). In this same chapter, Jesus also said: *"Take heed that ye despise not one of these little ones; for I say unto you, That in heaven their angels do always behold the face of my Father which is in heaven"* (Mat 18:10). Jesus said to take heed that we do not ignore even one child. Yet, many Christians look the other way when it comes to vaccines made from aborted fetal cell lines.

Maybe you have never heard these things. What is more shocking—*many have*. I could not believe there were Christians who didn't seem to care about this, or who weren't willing to investigate this little-known fact. Many cling to ignorance because they believe they must choose between their child who is alive, and someone else's child who is not.

Those who research this issue will come across the standard claims that two abortions were performed, these children "would have died anyway," and, of course, "God uses all things for good" (which is half of a verse taken out of its context—Satan's *modus operandi)*. Many Christians believe using vaccines is the only way to keep their children from dying of infectious diseases, and they find themselves trying to reconcile God's Word with this practice. The logical outworking of thinking it is acceptable to take the remains of "two children who would have died anyway" and use them to make fetal cell line vaccines and "save so many other children" is to believe God used the act of abortion to benefit mankind. With the wrong information, Biblical Christians come to the wrong conclusion. (Hab 1:4—*"the wicked doth compass about the righteous; therefore wrong judgment proceedeth."*) Satan deceives by appearing as an *"angel of light"* (2Co 11:14); he doesn't tempt us by showing us how evil something is, he tempts us by showing how "good" something evil can be made to look. Christians must not let fair words persuade us to allow heinous deeds.

Matthew 4:5 **Then the devil taketh him up into the holy city, and setteth him on a pinnacle of the temple,** 6 **And saith unto him, If thou be the Son of God, cast thyself down: for it is written, He shall give his angels charge concerning thee: and in their hands they shall bear thee up, lest at any time thou dash thy foot against a stone.** 7 **Jesus said unto him, It is written again, Thou shalt not tempt the Lord thy God.**

During Christ's temptation in the wilderness, Satan *quoted Scripture to Him.* Satan used God's words from Psalm 91:11-12 to try to tempt *the* Word *"made flesh"* (Jhn 1:14). Jesus then set an example for His church: He countered *Scripture taken out of its context* with *rightly divided* Scripture (2Ti 2:15). He showed us the two verses from the 91st Psalm do not negate the rest of the Bible; we must take all Scripture as a whole. Taking one verse out of its context to justify sin God condemns elsewhere is how Satan uses the Bible, it is not how Christians are to use the Bible. It is vital to keep passages in their place and to compare Scripture with Scripture to enlighten our understanding of the truth: the truth about salvation, doctrine, morality, and all other matters contained therein.

If using the Bible deceitfully is the way Satan chose to tempt the Messiah, we can be sure he uses this method on Christ's followers as well. Christians *need* to be fully aware that Satan is the god of this world (2Co 4:4) and *is in charge of it* (to a certain extent, for now—Rev 11:15); he uses the Bible deceitfully by taking verses out of their context. Satan is crafty and uses subtle half-truths to make us believe lies, all the while convincing us we are in the right by misusing Scripture to support our position. We need to recognize how susceptible we are to these methods. Satan deceives men by making them believe they are doing the exact opposite of what they are actually doing. These cell lines are no exception; God-fearing people are deceived into using aborted fetal cell line vaccines for their children's "health," but in reality, they are doing *harm.*

Satan tempted Eve by telling her she would be wise (Gen 3:5-6), not by telling her the terrible things that would happen if she didn't *completely* follow God's Word. Man fell because he was tempted with wisdom. Satan still tempts us this way, but true wisdom only comes from God; things that violate God's Word cannot be of His wisdom.

Proverbs 21:30 **There is no wisdom nor understanding nor counsel against the LORD.**

Jeremiah 8:9b ...**lo, they have rejected the word of the LORD; and what wisdom is in them?**

God *has told us* His Word can be used deceitfully (2Co 4:2). If any Christian thinks a Bible verse exists that can be used to justify abortion because it has been "used for good," they are deceived. It is improper to take a verse out of its context to support abortion: murder is condemned throughout the Bible. As the Bible declares in Job 14:4: *"Who can bring a clean thing out of an unclean? not one."*

Genesis 50:20 **But as for you, ye thought evil against me; but God meant it unto good, to bring to pass, as it is this day, to save much people alive.**

God *can* take absolutely horrible situations, work through obedient, God-fearing people who stay true to Him *despite the evil* and use them for good. That is what God does, not man. Christians are to follow God's Word and obey His commands to avoid even the *appearance* of evil (1Th 5:22). We are to trust God and live *within* the parameters of His laws (Psa 37:3). This passage is not permitting us to do something God calls evil. Christians are to live *according to Scripture* and not try to take matters into their own hands to accomplish what they think God ought. God will not use a practice that *should not exist* as the only way to keep our children from dying due to an infectious disease. God's way will never go against His Word. How could a God that delights in lovingkindness (Jer 9:24) have intended for us to murder babies to make medicine to keep other children alive? Christians are to follow His Word and *"overcome evil with good"* (Rom 12:21); we are not to acquiesce to evil because it's convenient. In the book of Genesis, we see Joseph as a type of Christ. He is betrayed and sold into slavery by his brothers, but God worked through this *obedient man* (Psa 105:19) in a horrible situation to save the Jewish people alive.

This was a special circumstance in Scripture: an Old Testament picture of the only death that *would* save mankind.

Romans 8:28 **And we know that all things work together for good to them that love God, to them who are the called according to his purpose.**

To use the first *half of* this verse to justify medical experiments on the remains of murdered children is to ignore much of God's revealed Word. Experiments on aborted children are unsurprising after effects of the devaluing of human life, which allowed their murders in the first place. All of which is completely against Scripture. The Bible is rife with the knowledge that human life was created by God and has much value to Him (Luk 12:7). Children are precious, and all human life is precious. Humans are not a medium on which to experiment for profit. While the reasoning for this "research" is supposedly to save lives, the reality is if there were no profit, few would be motivated to do this research.

"...not only has there never been a cure for any disease whatsoever using aborted fetal material, the true motive lies in the millions of dollars in federal grants that have been used for aborted fetal research."[4]

Yes, abortion is legal in this evil world, and maybe these children "would have died anyway"; this is not a license for Christians to participate in a practice that God *hates* (Jer 44:4).

Deuteronomy 12:31 **Thou shalt not do so unto the LORD thy God: for every abomination to the LORD, which he hateth, have they done unto their gods; for even their sons and their daughters they have burnt in the fire to their gods.**

We must compare Scripture with Scripture, rather than comparing Scripture with our experiences, emotions, or our very limited knowledge. To claim that God has used abortion for good is to speak wickedly and deceitfully for God (Job 13:7), and to change His judgments into wickedness (Eze 5:6). In Job 40:8, God said, *"Wilt thou also disannul my judgment? wilt thou condemn me, that thou mayest be righteous?"*

1 Samuel 2:3 **Talk no more so exceeding proudly; let not arrogancy come out of your mouth: for the LORD is a God of knowledge, and by him actions are weighed.**

Psalm 33:10 **The LORD bringeth the counsel of the heathen to nought: he maketh the devices of the people of none effect.** 11 **The counsel of the LORD standeth for ever, the thoughts of his heart to all generations.**

Believing these vaccines are beneficial is based on the false assumption that man can do something God says *cannot* be done. It would go *against His Word* to allow the act of abortion to be good for man, according to Matthew 7:18. Christ created all things (Heb 1:2), and by Him all things consist (Col 1:17). No amount of experimenting will accomplish what God will not allow—no matter how many children and animals are experimented on and no matter how many discoveries man makes. Christ upholds everything, every atom of creation (Heb 1:3). No interaction, reaction, realization, or ramification, great or small, will ever surprise Him. Man has never discovered any truth that contradicts God's Word, although otherwise claimed. The hand of the Lord has made everything in the earth, and in His hand is the soul of every living creature (Job 12:9-10). We cannot change what God's Word has established (Isa 14:27), and the world is established by *His* wisdom (Jer 10:12). Man's machinations do not void God's laws. A corrupt tree cannot bear good fruit (Mat 7:18), and the Scripture cannot be broken (Jhn 10:35).

How could murder bring forth anything but death? How is it Christians trust the words of those who are for abortion? Why would we entrust our children to a world system which believes it would have been acceptable to murder them before they were born, and then claims the children who are born need to be injected with aborted fetal cell line vaccines to keep them "healthy." God told Israel that His laws were for *their good*. Disobeying God's command to not kill will never result in anything that is beneficial for our children's health.

Deuteronomy 10:12 **And now, Israel, what doth the LORD thy God require of thee, but to fear the LORD thy God, to walk in all his ways, and to love him, and to serve the LORD thy God with all thy heart and**

with all thy soul, 13 **To keep the commandments of the LORD, and his statutes, which I command thee this day for thy good?**

Isaiah 45:19 **I have not spoken in secret, in a dark place of the earth: I said not unto the seed of Jacob, Seek ye me in vain: I the LORD speak righteousness, I declare things that are right.**

Christians are not to participate in murder, even for the supposed sake of "saving lives." Romans 3:8 declares that those who say, *"Let us do evil, that good may come,"* their *"damnation is just."* The "end justifies the means" is not a Biblical ideal if it means allowing sin in any way. This world makes merchandise of children; Christians are not to participate in this. We must let God declare what is right, not man. We must not let ourselves be hardened through the deceitfulness of sin (Heb 3:13), or let our own desires determine what we believe is true (Pro 18:1).

"Each medical benefit or scientific advance from the use of fetal tissue from elective abortions desensitizes beneficiaries, scientists, and doctors to the original evil act that produced these cells. Aborted fetal tissues used in laboratories are minimized to merely human cells, and the human beings whose lives were taken to provide those cells become irrelevant and with time forgotten. Of greatest concern is that desensitization ultimately leads to scandal by erroneously validating elective abortions for a greater good. Without careful oversight, the fetus could become, like fetal tissue cell lines, merely cells, cultured within the uterus for scientific exploration. All people of good conscience have the responsibility to voice opposition to the use of fetal tissue from elective abortions in order to promote development of alternatives, affirm the value of all human life, and limit scandal."[5]

A paper titled "Are There Really Alternatives to the Use of Fetal Tissue From Elective Abortions in Transplantation Research?" explains why the ethical alternatives are inferior to the preferred medium of an aborted baby. The study states: "Nevertheless the only advantage of using the suggested alternatives is the perception of them raising less ethical concern than fetal tissue from an induced abortion."[6] If something could only be accomplished by killing a child, it is not something that should be done.

Micah 3:2 **Who hate the good, and love the evil; who pluck off their skin from off them, and their flesh from off their bones;** 3 **Who also eat the flesh of my people, and flay their skin from off them; and they**

break their bones, and chop them in pieces, as for the pot, and as flesh within the caldron. 4 **Then shall they cry unto the LORD, but he will not hear them: he will even hide his face from them at that time, as they have behaved themselves ill in their doings.**

This passage isn't speaking about aborted fetal cell lines, but I believe it certainly applies. In an abortion, a child is killed, their remains are placed in a petri dish instead of a pot, and what is made from this experimentation is then "consumed" by injection. If putting pieces of a slain child in a petri dish is the only way certain kinds of *medicine* can *progress,* where else might this thinking lead? The Bible uses the word medicine twice, and neither instance has anything to do with killing children. Trust God—do not trust this world; making medicine should not involve murder. God warned us about *"science falsely so called"* (1Ti 6:20), and medicine is a *form of science.* Any "medicine" that requires killing unborn children is medicine, *falsely* so called.

Philippians 3:18 **(For many walk, of whom I have told you often, and now tell you even weeping, that they are the enemies of the cross of Christ:** 9 **Whose end is destruction, whose God is their belly, and whose glory is in their shame, who mind earthly things.)**

Jesus said that breaking one of the least commandments and *teaching others to do the same* would have significant consequences on our status in the kingdom of heaven (Mat 5:19). Is *"thou shalt not kill"* one of the *"least"* commandments? Jesus also told us it is not only wrong to kill, but it is also wrong to be angry without cause (Mat 5:21-22): how much more so is the murder of a child? Jesus was *"much displeased"* (Mar 10:14) when His disciples did not let children come to Him; how much more is He displeased with those who represent Him condoning and promoting the practice of murdering children for medical research? There is simply no Scriptural ground to stand on to support this wickedness. Jesus also said that which is highly esteemed among men is an abomination in the sight of God (Luk 16:15). Furthermore, Jesus warned about caring for our lives more than our souls (Mat 10:28; 16:26).

Acts 17:28 **For in him we live, and move, and have our being; as certain also of your own poets have said, For we are also his offspring.**

Philippians 4:19 **But my God shall supply all your need according to his riches in glory by Christ Jesus.**

God did not make any mistake in the design of our immune systems, much less one that can only be rectified by receiving vaccines made from aborted fetal cell lines. Did God not inspire David to write that man is fearfully and wonderfully made, even in his fallen state (Psa 139:14)? Why do God-fearing people believe God hasn't made provision for our children's immune systems (see Luk 11:40), and somehow abortions have accomplished this? Even though God specifically gave principles in His Word that prevent the spread of disease (which we will discuss soon), and He is the only one who heals? Christians must be more trusting of God and His Word. Sometimes the children of this world are wiser than those who say they believe every Word of God (Luk 16:8). Even some who believe in evolution do not allow that aborted fetal material can bring health. One author wrote the following about these cell lines:

"Sound a bit medieval? Well, maybe we can tell ourselves that hey, we're not all PhDs in biochemistry, so maybe there is some obscure scientific magic that happens when you culture virus on aborted fetal cells that will make infants healthier if you then inject it into their bloodstream, thereby offering them protection against a dread disease... Sounds logical."[7]

How any Christian can read the books of Isaiah, Jeremiah, or Ezekiel, and conclude that God has used abortion "for good" is mindboggling. In these Old Testament books detailing Israel's punishment for disobeying God's Word, the fact that they sacrificed their children to false gods is mentioned repeatedly. To take half of a verse completely out of its context to say God has used abortion for good is to pervert the words of the Living God (Jer 23:36). These books clearly draw a parallel between Israel's past judgment and Israel's future judgment, which will extend to the entire world (Lam 1:21-22). God's fury upon Israel's abominations is clearly explained. God said Israel had no right to question Him because they slew their own children (Eze 20:31). They went so far as to sacrifice their children and enter God's temple the same day (Eze 23:39). God has detailed in His Holy Word that killing children is unacceptable. Israel's past sin of sacrificing children correlates eerily to the same iniquity in our time in the form of abortion. (Ecc 3:15a: *"That which hath been is now."*) We might be fooled into believing God's long-suffering is apathy, but God cares very much about this matter. If you call yourself a Christian, it would be wise to have the same opinion of this issue that Christ does—you will have to answer to Him one day (2Co 5:10).

Ezekiel 16:20 **Moreover thou hast taken thy sons and thy daughters, whom thou hast borne unto me, and these hast thou sacrificed unto them to be devoured. Is this of thy whoredoms a small matter,** 21 **That thou hast slain my children, and delivered them to cause them to pass through the fire for them?**

The "greater good" logic which has brought much evil into this world is the same motivation behind those who sought Christ's death:

John 11:49 **And one of them, named Caiaphas, being the high priest that same year, said unto them, Ye know nothing at all,** 50 **Nor consider that it is expedient for us, that one man should die for the people, and that the whole nation perish not.** 51 **And this spake he not of himself: but being high priest that year, he prophesied that Jesus should die for that nation;**

Before you know it, the death of one man is not enough; they also considered killing Lazarus whom Christ had raised from the dead (Jhn 12:10-11). Christians should not use this same logic that resulted in the betrayal of the innocent blood (Mat 27:4) of Christ to convince us that the deaths of innocent children are acceptable to God!

Jesus said those who love Him will keep His words (Jhn 14:23). How have we come to the place where we believe using products made from aborted fetal cell lines is the only way to keep our children healthy? Is God so powerless? Is His Word so ineffectual? Do we really believe any good can come from killing children? Has God not told us murder is of the devil? The failure of Christians to believe and keep God's words has brought us to believe the words of those who promise life *through lies* (Eze 13:22). Why would Christians completely wrest the words "God uses all things for good" from their context and use them to justify their own desires (Pro 18:1) to use a product made from killing a child, when God Himself tells us He wouldn't even think of doing such a thing (Jer 7:31; 32:35)? A god that would require one child to die to save other children is a pathetic god; the God of the Bible is not so powerless (Rev 20:11). Why do we esteem vaccines so highly that we lower our estimation of God's holiness rather than question aborted fetal cell lines? It is imperative for us to realize there are two types of wisdom in this world: God's and Satan's.

James 3:15 **This wisdom descendeth not from above, but is earthly, sensual, devilish.** 16 **For where envying and strife is, there is confusion and every evil work.** 17 **But the wisdom that is from above is first pure, then peaceable, gentle, and easy to be intreated, full of mercy and good fruits, without partiality, and without hypocrisy.**

Satan, by his own wisdom accomplishes his goals by removing "*the bounds of the people*" (Isa 10:13). He appears as an angel of light (2Co 11:14), and convinces us to ignore God's laws. Aborted fetal medicine is not a manifestation of God using all things for good. This is not God's providence! Why do we believe God needs something as wicked as abortion to make life-saving medicine? God does not compromise with wickedness (Psa 94:20-21), and neither should professing Christians. Christians need the full counsel of God to discern good and evil (Heb 5:12-14). If we have not read, studied and believed the entirety of God's Word, we will fail to be wise considering that which is good, and *simple concerning evil* (Rom 16:19).

John 8:44 **Ye are of your father the devil, and the lusts of your father ye will do. He was a murderer from the beginning, and abode not in the truth, because there is no truth in him. When he speaketh a lie, he speaketh of his own: for he is a liar, and the father of it.**

Notice the connection between murder and lies. When Satan tempted Eve, most of what he said was technically true—with one glaring lie. But the impression he gave her about the consequences of her action was *far from the truth*. While there are many technically true statements about these aborted fetal cell lines, the overall narrative is false. We have been told two abortions were performed, making two cell lines, which is true. It is not claimed "only" or "just two" abortions were performed; these are my words. However, this true statement about two cell lines which have supposedly saved lives leads us to think in terms of lives saved versus lives lost, and we tend to use these words. Here is another example of what the public is told regarding aborted fetal cell lines from a web page titled "Vaccine Ingredients – Fetal Tissues" put forth by the Children's Hospital of Philadelphia:

"The fetal embryo fibroblast cells used to grow vaccine viruses were first obtained from elective termination of two pregnancies in the early 1960s. These same embryonic cells obtained from the early 1960s have continued to grow in the laboratory and are used to make vaccines today. No further sources of fetal cells are needed to make these vaccines."[8]

The picture painted by this true statement is also far from reality.

Chapter 2

Not Just Two, but Too Many to Count

"Can you understand that a small creature, who cannot even comprehend what is being done to her, in a vile place, in the dark and the cold, beats herself on her strained little chest with her tiny fist and weeps with her anguished, gentle, meek tears for 'dear God' to protect her—can you understand such nonsense, my friend and my brother, my godly and humble novice, can you understand why this nonsense is needed and created? Without it, they say, man could not even have lived on earth, for he would not have known good and evil. Who wants to know this damned good and evil at such a price? The whole world of knowledge is not worth the tears of that little child to 'dear God.' I'm not talking about the suffering of grown-ups, they ate the apple and to hell with them, let the devil take them all, but these little ones!"[1]

Isaiah 47:10 **For thou hast trusted in thy wickedness: thou hast said, None seeth me. Thy wisdom and thy knowledge, it hath perverted thee; and thou hast said in thine heart, I am, and none else beside me.**

God has given us a way to completely follow His laws without risking our children's health or lives, and we will get to that soon. First, let's seek the truth and *"Prove all things."* Let's see the reality behind the statement that there were two abortions performed, which provided material to make two cell lines. This true statement gives a false impression that somehow lulls Christians into accepting this practice. I hope the truth about the real number of abortions that took place to create these cell lines will startle these Christians out of their apathy.

The two abortions cited at the end of the previous chapter were used to make two aborted fetal cell lines called WI-38 and MRC-5:

"Cell lines derived from aborted fetal tissue have been fairly commonplace in research and medicine since the creation in the 1960s of the WI-38 cell strain, which was derived at the Wistar Institute in Philadelphia, Pennsylvania, and MRC-5, which came from a Medical Research Council laboratory in London. Viruses multiply readily in these cells, and they are used to manufacture many globally important vaccines, including those against measles, rubella, rabies, chicken pox, shingles and hepatitis A. An estimated 5.8 billion people have received vaccines made with these two cell lines…"[2]

It has been briefly established that two abortions took place to make two cell lines, called WI-38 and MRC-5. These facts are not disputed, and there will be no more time dedicated to prove what is commonly understood. Now, let's look at the evidence that gives us a true measure of the number of abortions which took place for vaccine research, not just the two cell lines used.

"In 1975, the Coriell Institute for Medical Research, under contract to the National IMR-90 Institute on Aging (NIA) to establish and maintain an Aging Cell Repository, was directed to establish and bank for future research a new human diploid cell line as a replacement for the cell line known as WI-38. WI-38 had been established and characterized in the early 1960's as a standard (reference cell line) in the burgeoning field of cell biology; but in the early 70's, the NIH stock of low passage WI-38 cells had become seriously depleted. The cell line developed at Coriell, identified as IMR-90 (1), was the first of several lines planned in support of NIA research programs and general cell biology research. … The IMR-90 cell line, like WI-38, was derived from lung tissue of a human female embryo following therapeutic abortion."[3]

"Therapeutic abortion," an idiotic term.

Here is another cell line that looks as though it is being tested to see if it could replace MRC-5. This study shows us different cell lines have idiosyncratic behaviors, some replicating faster than others.

"Human diploid cell strains (HDCSs), possessing identical chromosome sets known to be free of all known adventitious agents, are of great use in developing human vaccines. However it is extremely difficult to obtain qualified HDCSs that can satisfy the requirements for the mass production of vaccines. We have developed a new HDCS, Walvax-2, which we derived from the lung tissue of a 3-month-old fetus. … Specifically, Walvax-2 cells replicated more rapidly than MRC-5 cells, with Walvax-2 cells attaining the same degree of confluence in 48 hours as was reached by MRC-5 cells in 72 hours. Moreover, Walvax-2 cells attained 58 passages of cell doublings whereas MRC-5 reached 48 passages during this period. We also assessed the susceptibility of these cells to rabies, hepatitis A, and Varicella viruses. Analysis of virus titers showed the Walvax-2 cells to be equal or superior to MRC-5 cells for cultivating these viruses. … In conclusion, results from this study show that Walvax-

2 cell banks are a promising cell substrate and could potentially be used for the manufacturing of HDCVs [Human Diploid Cell Vaccines]."[4]

With all of the abortions happening throughout the world, do we really believe "better" cell lines are not continually trying to be made?

There are other cell lines besides WI-38 and MRC-5, such as IMR-90, Walvax 2 (both listed above), WI-26,[5] HEK293,[6] PER.C6,[7] IMR-91,[8] WI-44,[9] and others.[10] Whether or not these other cell lines have, in fact, been utilized to make vaccines used on the public is not something we will delve into, as it is difficult to discern the truth when it comes to this subject. While we can see the cell line IMR-90 was made as a replacement for WI-38 because its supply was low, there does not appear to be easily verifiable evidence of cell lines other than WI-38 or MRC-5 being used for vaccines on the CDC's immunization schedule.

Why a new cell line was made as a replacement for WI-38 yet supposedly not used as a replacement for this older cell line is uncertain. It has been said to get to the truth of certain issues surrounding immunization "...generally calls for the resources of something like the secret service."[11] I only want to present the undeniable facts and I am without these resources. These cell lines obviously exist and are being used for medical research, which medical literature can verify. An organization called *Children of God for Life* has a vaccine chart on its website, which shows some of these other cell lines being used to make vaccines that are not on the CDC's schedule (see this link in the additional resources). My main point is to clarify misconceptions regarding the statements made about the two abortions, which provided the material for two aborted fetal cell lines.

The assertion that all vaccines were made from two cell lines which came from two abortions and no new sources are necessary, appears to be untrue simply because of the creation of the IMR-90 cell line as a replacement for a diminishing supply of WI-38 "stock." In the past, tens of thousands of monkeys were killed to make polio vaccines: "...the large number of monkeys (about 1500) needed to be sacrificed to produce every 1 million inactivated doses."[12] *Sacrificed*. One thousand, five-hundred monkeys per one million doses. Yet we are to believe two abortions provided all of the material necessary to make five different vaccines, not to mention the different versions made by different companies for nearly 60 years? Technology has obviously advanced since the development of the polio vaccine, but I simply don't believe the public is told the whole story.

It is hard to know exactly why new cell lines were made, but supposedly never used. Instead of looking at what is happening now, let's look at some older information that is undisputed. I believe looking at what was verified in the past gives us a better measure of what is happening today. An abundance of evidence shows a large number of experiments on aborted children *before* the successful creation of a cell line. *Children of God for Life* gives information showing one example of the number of abortions not reported. Their website tells us: "67 abortions to produce rubella virus plus 32 abortions to produce the cell line for cultivation means there was a total of **at least 99 elective abortions to create the rubella vaccine alone.**"[13] and Dr. Stanley Plotkin had an "open testimony on [the] use of 76 aborted fetuses."[14]

Before a cell line is created, the experiments are denoted by the numbers in the cell line names: WI-38, MRC-5, WI-44, HEK293, etc. Each experiment is considered a cell strain, and when the scientists have worked their craft, they can make a cell strain into a cell line. A cell line is considered immortal, while a cell strain is not. A paper published in *Experimental Cell Research* in 1961 with the aim to "...describe and characterize the development of 25 strains of human cells derived from fetal tissue..."[15] has a table showing 19 fetuses were used to make 25 cell strains: WI-1 through WI-25.[16]

The public has been told about the two cell lines, but the experimentation with cell strains required to make the cell lines is not openly disclosed. While it cannot, as far as I know, be shown that these strains have been used to make any vaccines that are on the CDC's immunization schedule, they were certainly used in the preparatory vaccine research that allowed the two cell lines WI-38 and MRC-5 to be made. The same paper states cell "lines" at the beginning of the study but goes on to say they are actually cell "strains."

"These cell lines [strains], of which over two hundred have been reported in the literature, have serious limitations for many kinds of biological studies."[17]

In 1961 there were over 200 cell strains. This study also tells of the "economic advantages" of using a cell strain.

"The isolation and characterization of human diploid cell strains from fetal tissue make this type of cell available as a substrate for the production of live virus vaccines. Other than the economical advantages, such strains in contrast to heteroploid cell lines..."[18]

Also, even though this study is about human fetal cells, adult cells have also been tested, and basically do the same thing.

"Although the subject of this report is confined to experiments involving human fetal cells, adult human cells have also been carried for similarly extensive periods of time with retention of the diploid configuration. Other workers have reported similar results with adult human diploid cells."[19]

Many Christians use the arguments from Chapter 1 to speak on this issue when in fact they have not devoted much, if any, time to researching this topic. I want to give these Christians a wake-up call and show them what they are defending. While I have not included anything terribly graphic, the next few pages contain accounts that are not pleasant.

In the previously cited study, we are told about some of the methods used.

"The fragmented lungs, for example, from a three-month-old human fetus were usually placed in four Blake bottles."[20]

Here are more examples showing the high number of aborted children for these experiments.

"Human diploid cell strains.—Strains WI-26, IVI-38 and WI-44 were used. WI-26 was derived from male fetal human lung and WI-38 and WI-44 from female human fetal lung. All embryos were obtained from surgical abortions and were of approximately three months' gestation."[21]

"Explant cultures were made of the dissected organs of a particular fetus aborted because of rubella, the 27th in our series of fetuses aborted during the 1964 epidemic."[22]

"Two groups of human fetuses 8-20 weeks of age were used for the initiation of diploid cell strains. The first group consisted of normal embryos obtained by hysterotomy and flown from Scandinavia, bathed in physiologic saline at 0° C. The second group represented spontaneous abortions obtained from the gynecologic service of the Philadelphia General Hospital. Intact fetuses of both groups were kept at 0-4° C from the time of abortion until used, 12 to 96 hours later. ... Cell strains derived from 29 fetuses were examined. The following tissues and organs yielded viable cell strains: skin, lung, pharyngeal mucosa, pericardium, brain, pituitary, kidney and thymus. ... Sixtythree [sic] cell strains, derived from 29 fetuses by means of an organ culture technique, were studied."[23]

Notice the callous disregard for human life.

"Cell strains derived from 29 fetuses were examined. Twentyone [sic] of these originated from surgical abortions, while 7 came from spontaneously aborted fetuses. One cell strain was of uncertain origin. At the start of these studies, most importance was attached to HDCS derived from the surgically aborted fetuses since these could be presumed to be normal. In fact, no differences in any of the parameters studied could be found between the two groups of fetuses, and no distinction will henceforth be made between them."[24]

"The stability and integrity of the human foetal cell strain WI-38, which has been well demonstrated in the past ten years, and its susceptibility to viruses infective for man, explain the value of such material for the isolation of viruses and in the development of vaccines. We have developed another strain of cells, also derived from foetal lung tissue, taken from a 14-week male foetus removed for psychiatric reasons from a 27 year old woman with a genetically normal family history and no sign of neoplastic disease both at abortion and for at least three years afterwards."[25]

"The particular advantage of this technique for the present studies is that it enables cultures to be established from exceedingly small amounts of tissue, an obvious advantage when working with organs derived from [immature] fetuses."[26]

"Source of Virus: Virus was obtained from an aborted rubella-infected human fetus. The 25-year-old mother was exposed to rubella... The fetus was surgically [aborted] 17 days after the maternal illness and dissected immediately."[27]

"Dr. K McCarthy: It seems to me that there are two things that we worry about in regard to WI-38 cell substrate. First of all, presence of extraneous viral agents; secondly, the possibility of there being human genetic material passed over into the vaccine. I wonder if there is any information about the reasons for aborting that particular embryo that gave rise to WI-38; and if it was from a family, whether we have any information about siblings from the family and whether they are normal?

Dr. S. Plotkin, Philadelphia: I should like to answer Dr. McCarthy's question. This fetus was chosen by Dr. Sven Gard, specifically for this purpose. Both parents are known, and unfortunately for the story, they are married to each other, still alive and well, and living in Stockholm, presumably. The abortion was done because they felt they had too many children. There were no familial diseases in the history of either parent, and no history of cancer specifically in the families; that is, the maternal or paternal sides. I believe this answers Dr. McCarthy's question."[28]

The fact that myself and nearly every person who reads this material, along with billions of others worldwide, have all been injected with the DNA of the unwanted child in the above quote is incredibly disturbing, to say the least.

Many of these studies were co-authored by a man named Stanley Plotkin. He willingly testified in a court case between two parents who had different opinions regarding vaccination. Here is a revealing excerpt from his deposition.

"**QUESTION:** In your work related to vaccines, how many fetuses have been part of that work?

PLOTKIN: My own personal work? Two.

QUESTION: So in your, in all of your work related to vaccines throughout your whole career, you've only ever worked with two fetuses?

PLOTKIN: In terms of making vaccines, yes.

QUESTION: (Presented with Exhibit 41). Are you familiar with this article, [Dr. Plotkin]?

PLOTKIN: Yes.

QUESTION: Are you listed as an author on this article?

PLOTKIN: Yes.

[Some questions are skipped.]

QUESTION: So this was related to your work?

PLOTKIN: Well, yes, in a sense.

QUESTION: To vaccines, correct?

PLOTKIN: <u>Yes. It was preparatory</u>.

QUESTION: So this study involved 74 fetuses, correct?

[Some questions are skipped.]

PLOTKIN: Seventy-six.

QUESTION: And these fetuses were all three months or older when aborted, correct?

PLOTKIN: Yes.

QUESTION: And these were all normally developed fetuses, correct?

PLOTKIN: Yes.

QUESTION: These included fetuses that were aborted for social and psychiatric reasons, correct?

PLOTKIN: Correct.

QUESTION: What organs did you harvest from these fetuses?

PLOTKIN: Well, I didn't personally harvest any, but a whole range of tissues were harvested by co-workers.

QUESTION: And these pieces were then cut up into little pieces, right?

PLOTKIN: Yes.

QUESTION: And they were cultured?

PLOTKIN: Yes.

QUESTION: Some of the pieces of the fetuses were pituitary gland that were chopped up into pieces.

PLOTKIN: Mm-hmm.

QUESTION: Included the lung, skin, kidney, spleen, heart and tongue of the fetuses?

PLOTKIN: Yes.

QUESTION: So I just want to make sure I understand. In your entire career -- this was just one study. So I'm going to ask you again, in your entire career, how many fetuses have you worked with approximately?

PLOTKIN: Well, I don't remember the exact number, but quite a few when we were studying them originally before we decided to use them to make vaccines.

[Some questions are skipped.]

QUESTION: Have you ever used orphans to study an experimental vaccine?

PLOTKIN: Yes.

QUESTION: Have you ever used the mentally handicapped to study an experimental vaccine?

PLOTKIN: [I don't recollect ever doing studies in mentally handicapped individuals. At the time, in the 1960s, it was not an uncommon practice.]

QUESTION: There's an article entitled "Attenuation of RA 27/3 Rubella Virus in WI-38 Human Diploid Cells." Are you familiar with that article?

PLOTKIN: Yes.

QUESTION: In that article, one of the things it says is 13 -- is one of the things it says is: 13 seronegative mentally retarded children were given RA 27/3 vaccine?

PLOTKIN: Okay. Well, then that's, in that case that's what I did.

[Some questions are skipped.]

QUESTION: Have you ever used babies of mothers in prison to study an experimental vaccine?

PLOTKIN: Yes.

QUESTION: Have you ever used individuals under colonial rule to study an experimental vaccine?

PLOTKIN: Yes.

QUESTION: Did you do so in the Belgian Congo?

PLOTKIN: Yes.

QUESTION: Did that experiment involve almost a million people?

PLOTKIN: Well -- well, all right, yes."[29]

(You can also see this interview on YouTube.[30,31])

Psalm 119:53 **Horror hath taken hold upon me because of the wicked that forsake thy law.**

1 Timothy 6:20 **O Timothy, keep that which is committed to thy trust, avoiding profane and vain babblings, and oppositions of science falsely so called:** 21 **Which some professing have erred concerning the faith. Grace be with thee. Amen.**

This is not science, it is death profiteering, and the mass murders taking place in the form of "therapeutic" abortions are a holocaust. Even if vaccines had no association with abortion, I would still want nothing to do with a product tested on orphans, mentally handicapped children, and helpless babies born to incarcerated women. It shouldn't be surprising that those who support abortion also have no problem with experimenting on other human beings. Violating God's laws leads to further wickedness. These tests for the sake of experimental vaccines are *crimes against humanity*. Mind you, this man is not a Nazi war criminal, but a venerated scientist who helped create the rubella vaccine and his book *Vaccines* is the standard reference for physicians. He is praised for *helping* children by developing vaccines that have "saved lives." According to God's Word, it is not possible for this wickedness to accomplish anything good (Mat 7:18). Thank the Lord that justice will be done, and these people will be held responsible for their acts (Rom 1:18). It would be better for all men to turn from their wickedness (Eze 18:23-32; 33:11) and come to Christ, as He wills (2Pe 3:9); but we know there will be men who will not repent of their murders, nor of their *sorceries* (Rev 9:21).

I would not want anyone performing these studies to even be in the same room as my child; I'm not going to *inject her* with the products made from these horrifying experiments. If you are a true Bible-believing Christian, then your beliefs are *incompatible* with the beliefs of those who conduct these experiments. Those who do not know God are wise to do *evil* (Jer 4:22). In the book of Exodus, it was a *fear of God* that kept the midwives from killing the Hebrew children (Exo 1:16-17). Those whose feet are swift to shed blood have no fear of God before their eyes (Rom 3:15; 18). True Christians are abominations to people like Stanley Plotkin (Pro 29:27), and if you think they value your children's lives, you're sadly mistaken. I'm sure many of them are under some greater good delusion, but what is good to these people? Aborted fetal experiments are apparently good; what else? Population control? Eugenics? Survival of the fittest? What will they do when your child is in the way of their so-called greater good? It is clear from Scripture: these abortions and experiments are wicked. The Israelites were told that those who took payment to kill the innocent are *cursed*. How could we believe this would ever bring our children health?

Deuteronomy 27:25 **Cursed be he that taketh reward to slay an innocent person. And all the people shall say, Amen.**

Fellow Christians, if you were under the impression that a purported "small" number of abortions have been used "for good" because they were used to make vaccines, I hope you are now of a different opinion.

Isaiah 5:20 **Woe unto them that call evil good, and good evil; that put darkness for light, and light for darkness; that put bitter for sweet, and sweet for bitter!**

Galatians 3:1a **O foolish Galatians, who hath bewitched you, that ye should not obey the truth…?**

As Paul questioned, who has beguiled you to ignore God's clear command for man's wisdom? This deception has led to the death of children—the same result Eve experienced after disobeying God's words (Gen 4:8). The technically true statement about these cell lines gives an impression that is untrue; there were not "just two" abortions. The real number is shown to be much higher based on the additional cell lines, the number of babies experimented on to create a cell line, and the experimenting to find better cell lines. Another article published by *Children of God for Life* titled "Forsaking God for the Sake of Science" shows the extent of this wickedness justified in the name of "Science." It is not for the fainthearted, but if you have knowingly received vaccines using an aborted fetal cell line, I believe you owe it to those who have suffered to read this article. Their murders *should* make us uncomfortable. Thank God that these innocent children are with Him (2Sa 12:23).

We have only looked at a few verses condemning murder, but more are found throughout the Bible. The Scriptures warn about establishing upon unrighteousness (Jer 22:13); Psalm 127:1 says except the LORD build a house they labor in vain that build it. Here are some verses that say God will repay man according to his doings: Psa 94:23; Eze 11:21; Jer 21:14; Dan 9:13. These verses tell us that God will repay bloodshed: Eze 9:9-10, 20:26, 22:13, 35:6, and 36:18. God repeatedly judged Israel for violating the law they had agreed to keep, and that day is coming when the entire world will be judged.

Habakkuk 2:12 **Woe to him that buildeth a town with blood, and stablisheth a city by iniquity!**

Even *if* it was possible for murder to save lives, murdering even one child to save the entire world is evil before God. The only exception is when God sacrificed *his own Son* for the sins of man. God made Jesus "*...to be sin for us, who knew no sin; that we might be made the righteousness of God in him*" (2Co 5:21). Repenting (Act 20:21) and believing the gospel (1Co 15:3-4) is the only way for man to escape the eternal consequences of sin. Repenting means to change your mind: change your mind about what you believe about your sin and your salvation to what God says about these things. Change from trusting yourself: your belief that you are a "good person," your religion, your good works, your baptism, or church membership, and instead trust what Christ accomplished on the cross (Luk 9:31).

John 6:29 **Jesus answered and said unto them, <u>This is the work of God,</u> that ye <u>believe</u> on him whom he hath sent.**

Acts 4:12 **Neither is there salvation in any other: for there is none other name under heaven given among men, whereby we must be saved.**

As Jesus said: *"repent ye, and believe the gospel"* (Mar 1:15). God's Word declares in Romans 10:9, *"That if thou shalt confess with thy mouth the Lord Jesus, and shalt believe in thine heart that God hath raised him from the dead, thou shalt be saved."* Jesus is the <u>only way</u> (Jhn 14:6) to eternal life; if salvation could be earned by good works, then *"...Christ is dead in vain"* (Gal 2:21). Jesus is the only way, and His laying down (Jhn 10:18) of His own life was the only way that man could be redeemed. Jesus said, *"Greater love hath no man than this, that a man lay down his life for his friends"* (Jhn 15:13). Jesus sacrificed himself, which was an act of great love; He has never asked us to sacrifice our children because this is an act of great evil. Child sacrifice is a sign of a false and wicked belief system.

Would a completely holy God who hates murder leave us—*His children*—with a choice between our children's lives and the lives of other children? Would the only way to be healthy involve a wicked practice that God, the only One who can heal, hates? Is there really only one way for a child to survive illness—by another child dying to supply the cells needed to make a vaccine and millions of helpless infants and unsuspecting people being experimented on for vaccine studies? The answer is no. The necessity of this kind of "medicine" is a lie. The statements about aborted fetal cell lines have been proven misleading: there have been terrible actions committed to make vaccines for the "good" of society. Should the people trust that we have been told the plain truth about the necessity of these vaccines in the first place? Man's way is not the only way. The best methods to avoid disease were given to man thousands of years ago, in God's Word.

Colossians 2:8 **Beware lest any man spoil you through philosophy and vain deceit, after the tradition of men, after the rudiments of the world, and not after Christ.**

Colossians 2:3 **In whom are hid <u>all</u> the treasures of wisdom and knowledge.**

Chapter 3

God's Way to Avoid Disease

Isaiah 46:9 **Remember the former things of old: for I am God, and there is none else; I am God, and there is none like me, 10 Declaring the end from the beginning, and from ancient times the things that are not yet done, saying, My counsel shall stand, and I will do all my pleasure:**

Jeremiah 10:23 **O LORD, I know that the way of man is not in himself: it is not in man that walketh to direct his steps.**

Proverbs 19:21 tells us there are many devices in a man's heart, but only the counsel of the LORD will stand. Psalm 94:10 reasons that the Author of knowledge must know all things. David wrote in Psalm 119:99-100 that he had more understanding than his teachers and the ancients because he kept God's precepts. Jesus said in Luke 16:17, "*...it is easier for heaven and earth to pass, than one tittle of the law to fail.*" There are "*wondrous things*" in God's Law (Psa 119:18). When God spoke to Moses on Mount Sinai, He gave "*lively oracles*" (Act 7:38), "*right judgments, and true laws, good statutes and commandments*" (Neh 9:13).

Exodus 15:26 **And said, If thou wilt diligently hearken to the voice of the LORD thy God, and wilt do that which is right in his sight, and wilt give ear to his commandments, and keep all his statutes, I will put none of these diseases upon thee, which I have brought upon the Egyptians: for I am the LORD that healeth thee.**

Deuteronomy 4:6 **Keep therefore and do them; for this is your wisdom and your understanding in the sight of the nations, which shall hear all these statutes, and say, Surely this great nation is a wise and understanding people.**

God told the Israelites that His commandments would keep them from disease. The commandments He gave dealt with righteousness, cleanliness, and diet (Heb 9:10). There will be no discussion about diseases caused by immoral behavior, aside from choosing to use aborted fetal cell line vaccines, but we will see that cleanliness and nutrition have much to do with preventing many infectious diseases. The last two verses are not used to imply the Church has any claim to the promises of Israel, but that there are laws which God has established with consequences that affect all men. In the books penned by Moses (2Ch 34:14; Jhn 5:46; Luk 16:31; 20:28), God gave His people instructions to prevent infectious diseases.

God commanded the Israelites to quarantine those who were sick (Lev 13:46), to "*...cover that which cometh from thee...*" (Deu 23:13), to wash with *running* water (Lev 15:13; Num 19:17), and even to *cover their mouths* (Lev 13:45). These are all crucial methods to prevent spreading disease. Also, God could have told them to utilize ancient inoculation methods but didn't. God's Word is full of scientific truths that have repeatedly proven superior to man's wisdom.[1] God's laws were a revelation to His people containing simple rules about sanitation. Throughout history, man's ignorance of these laws has led to disease and death in unimaginable proportions.

Deuteronomy 29:29 **The secret things belong unto the LORD our God: but those things <u>which are revealed</u> belong unto us and to our children for ever, that we may do all the words of this law.**

God also told His people that sickness would be one of many consequences if they did not obey His laws.

Deuteronomy 28:15 **But it shall come to pass, if thou wilt not hearken unto the voice of the LORD thy God, to observe to do all his commandments and his statutes which I command thee this day; that all these curses shall come upon thee, and overtake thee: ... 22 The**

LORD shall smite thee with a consumption, and with a fever, and with an inflammation, and with an extreme burning…

(See also Lev 26:14-16.)

Deuteronomy 6:24 **And the LORD commanded us to do all these statutes, to fear the LORD our God, for our good always, that he might preserve us alive, as it is at this day.**

The Jews were supernaturally blessed or cursed for obeying or disobeying God's laws. The sickness resulting from the improper disposal of human waste does not fall under the category of a supernatural repercussion that only applies to the Jewish people—this is obviously a natural law that affects all men. Vaccines did not cause infectious disease to diminish; God's laws of sanitation did, but man has given his own invention the credit. Here is some secular knowledge showing that vaccines are not responsible for reducing infectious diseases, starting with a paper published in *Pediatrics,* the official journal of the American Academy of Pediatrics.

"Once again, nearly 90% of the decline in infectious disease mortality among US children occurred before 1940, when few antibiotics or vaccines were available. … The major declines in child mortality that occurred in the first third of the 20th century have been attributable to a combination of improved socioeconomic conditions in this country and the public health strategies to protect the health of Americans. These public health measures…including water treatment, food safety, organized solid waste disposal, and public education about hygienic practices. … Vaccination, while first used in the 18th century, became more widely implemented in the middle part of the century. Vaccines against diphtheria, tetanus, and pertussis became available during the late 1920s but only widely used in routine pediatric practice after World War II. Thus vaccination does not account for the impressive declines in mortality seen in the first half of the century."[2]

"Are vaccinations primarily responsible for the decline of infectious diseases? Actual statistics and records from around the world show that infectious diseases—for example, smallpox, diphtheria, whooping cough, and scarlet fever—began to disappear long before immunizations ever came on the scene. According to the *World Health Statistics Annual*, 1973-1976, Vol. 2, there has been a steady decline of infectious diseases 'In most 'developing' countries regardless of the percentage of immunizations administered in these countries. It appears that generally improved conditions of sanitation are largely responsible for preventing 'infectious' diseases."[3]

"Health authorities credit vaccines for disease declines, and assure us of their safety and effectiveness. Yet these assumptions are directly contradicted by government statistics, published medical studies, Food and Drug Administration (FDA) and Centers for Disease Control (CDC) reports, and the opinions of credible research scientists from around the world. In fact, infectious diseases declined steadily for decades prior to mass immunizations…"[4]

"There is no convincing scientific evidence that mass inoculations can be credited with eliminating any childhood disease."[5] Robert Mendelsohn, MD

"An extensive study of medical literature reveals that there is no evidence whatsoever of the ability of vaccines to prevent any diseases. To the contrary, there is a great wealth of evidence (direct and indirect) that they cause serious side-effects."[6] Viera Scheibner, PhD

"It is not strange that health improves when the population gives up using diluted sewage as the principle beverage."[7]

"Infectious disease mortality had a precipitous and sustained decline during the entire twentieth century. Before the advent of federal vaccination policy, health officials attributed these gains to the significant investments the country made in modern sanitation and hygiene infrastructure. … Infectious diseases were on their way out before vaccination became commonplace. Some diseases for which vaccines were never developed or routinely administered—such as typhoid fever, scarlet fever, and the bubonic plague—have almost completely disappeared. Public health officials simply do not know what would have happened in the absence of widespread vaccination. They cannot assert with certainty that vaccines are responsible for disease eradication or that vaccines are required to maintain herd immunity."[8]

Here is a quick explanation of the theory behind herd immunity:

"What is salient and troubling about the theory of herd immunity is that the original basis for its validity has nothing to do with vaccines. People observed a protective effect in the community when a sufficiently high number of individuals contracted the wild form of a disease and secured lifelong immunity. This enduring, cumulative protection from the natural acquisition and resolution of disease is what public health officials called 'herd immunity.' … Vaccine-induced immunity, however, is not the same as natural immunity. A person who contracts measles in early childhood will never get the disease again; a person with vaccine-induced immunity will need to receive boosters or assess his or her antibody levels periodically. Vaccine-induced immunity is qualitatively different. Vaccines do not always work, and even when they do, their protection weakens over time. There are many examples that throw the validity of herd immunity into question. Outbreaks of diseases such as measles, mumps, and chickenpox, among others, routinely occur in fully vaccinated communities."[9]

Many mortalities from communicable diseases had already decreased substantially before their respective vaccines were introduced. (Measles mortality declined 99.96% from its peak in

England.[10]) As this unknown truth shows, the real reason for this decrease was clean water, proper disposal of human waste, and a better understanding of sanitation, food safety, and improved nutrition. This is probably hard to accept for those who have been told the opposite their whole lives, but this *aligns* with what God's Word declares about avoiding disease.

Please don't take my word for it; I greatly encourage you to research this for yourself. Beware of authors stating figures that show a marked reduction in illness by defining the "pre-vaccine" era as being *right before* these vaccines were widely used on the population—when in fact these diseases had already been greatly reduced. One page of statistics from Our World in Data shows that measles mortality has declined 100% since the measles vaccine was used on the public. But if you look closely, it states the pre-vaccine era of measles as 1953-62 with 2.5 deaths per million per year.[11] Most just see the 100% reduction and look no further.

"Permitting statistical treatment and the hypnotic presence of numbers and decimal points to befog causal relationships is little better than superstition. And it is often more seriously misleading."[12]

Also notice the use of the terms incidence vs. mortality. The incidence of some infectious diseases have reduced since the institution of mass vaccinations. Vaccines have prevented some **occurrences** of disease, but not without many dire consequences, as we will see. But they have not had much, if any, effect on the **mortalities** from these diseases. The mortality rate, however, is a more accurate measure of infectious diseases.

"It is an interesting fact that the death rate or number of deaths often is a better measure of the incidence of an ailment than direct incidence figures—simply because the quality of reporting and record-keeping is so much higher on fatalities."[13]

Watch for those who parrot the benefits of vaccines without giving any data to support their statements. Some authors make claims about the incidence and mortality of diseases, and the effectiveness of their corresponding vaccines that simply do not align with the data from the time period on which they are speaking. Dr. Suzanne Humphries, in her book *Dissolving Illusions: Disease, Vaccines and the Forgotten History,* compares some of these statements and the historical data, showing that these remarks are misleading.[14] She sums up in her book:

> "We are not saying that vaccines had absolutely no impact on deaths from these infectious diseases, but by using historical statistics, it is clear that most authors' claims regarding the lifesaving effects of vaccination are markedly exaggerated, and **risk is continuously denied or underplayed**. Despite the fact that 'vaccines were responsible for the massive decline in deaths' is based on a false foundation, it permeates medical thinking all over the world. Statements found in medical literature uniformly exaggerate the benefit of vaccination."[15]

Viera Scheibner, PhD, after studying thousands of pages of medical literature, wrote: "Reports of increased epidemics shortly after a fall in vaccination compliance are quite untrue and, at best, exaggerated."[16] (She was speaking specifically about pertussis.)

Sources that state epidemics following periods when there were lower vaccination rates—*if* these statements are true, we will see much evidence to the contrary—have underpinnings which are highly debatable assumptions: that vaccination would have absolutely saved any lives lost, and that the negative reactions caused by the vaccines would cause less damage than the disease. (These assumptions also disregard the facts that there are other ways of preventing and treating disease.) For example, when Japan postponed their vaccination age, their SIDS deaths "disappeared," according to a report published in *Pediatrics*. (We will discuss SIDS in the next chapter.)

> "The category 'sudden death' is also instructive in that the entity disappeared following both whole-cell and acellular vaccines, when immunization was delayed until a child was 24 months of age."[17]

If one were to read of a rise in pertussis following decreased vaccination rates, they would probably shake their heads at the parents who chose not to get this vaccine for their children. However, when you see that a change in vaccination practice caused SIDS to "disappear" in a developed country, this changes one's reaction. Make sure that no matter what you read, you look at all of the facts, not just one snippet of information taken out of the context—especially if that information is coming from someone who makes a living selling vaccines.

Next, smallpox and polio will be briefly examined, as their "disappearances because of vaccines" are deeply entrenched beliefs.

Smallpox

"In matters of science and the healing-art, there is no such thing as fact absolutely established beyond future investigation. Every position has its beginning from an anterior supposition, and may be superseded by later discovery."[18]

In general, researching vaccines is a daunting task, and it is hard to know what is true when conflicting information abounds. I have always supported a parent's right to choose whatever they want for their child, but I also thought it was foolish to reject vaccines. When I really began to research this subject, I had a newborn baby girl who was entirely dependent on me. My preconceived notions were not important compared to being absolutely sure something was safe for my child. It really is hard work to get to the truth regarding vaccines, and what follows will conflict with what you have always been told. But I firmly believe it is the truth.

"To take issue with a popular idea requires moral courage transcending any common audacity. Only the conviction of being right enables one to speak when thus confronted. In the medical profession especially, innovation upon accepted methods is bitterly resented. New remedies, new modes of treating the sick, and new ideas of pathology, are alike unwelcome. They disturb the sleep of ages, and show men faulty who had been regarded as well nigh infallible. Such offences are rarely condoned. The person committing them receives generally proscription and abuse, but rarely a candid hearing."[19]

Any time vaccines are questioned, the questioner or their information will be labeled as "unscientific." We must realize that not investigating is unscientific, not examining is unscientific, not questioning is unscientific. Letting others tell you what to think and not giving factual reasons to show how you *know* something to be true is unscientific. Furthermore, believing in something your whole life without looking at the facts for yourself is unscientific. Please follow the Bible, and hear the matter before you answer (Pro 18:13). Please be scientific and do not condemn before you investigate. I don't expect everyone to come to all the same conclusions that I do, but I do expect others to know *why* they believe what they believe. These accounts, most of which are from doctors practicing in the late 1800 to early 1900s, led me not to believe the oft-quoted line that vaccination eradicated smallpox.

"With the present century vaccination has become the hobby rather than the conviction of medical men. It is accepted, with almost an unreasoning deference to the opinions of others; indifference to any intelligent view of the subject, and a disposition to follow in the groove established by precedent, without a thought whether that precedent is right or wrong, rational or fallacious. ... Our veneration for science and truth is deep and broad; but no dogma, in whatever department of thought has even the commonest title to our respect if it cannot show some other title except the antiquity which invests it with mystery, or the imposing names which give it currency."[20]

Let's examine the historical facts about smallpox and see if the common understanding of this illness and its disappearance aligns with these facts. In a country where the religion is "Science," with vaccines being the crowning achievement, this is a dogma not easily denounced.

"From childhood I have been trained to look upon the cow-pox [smallpox vaccination] as an absolute and unqualified protective. I have, from my earliest remembrance, believed in it more strongly than in any clerical tenet or ecclesiastical dogma. The numerous acknowledged failures did not shake my faith. I attributed them either to the carelessness of the operator or the badness of the lymph. In the course of time, the question of vaccine compulsion came before the Reichstag, when a medical friend supplied me with a mass of statistics favorable to vaccination, in his opinion conclusive and unanswerable. This awoke the statistician within me. On inspection, I found the figures were delusive; and a closer examination left no shadow of doubt in my mind that the so-called statistical array of proof was complete failure."[21]

I compiled the following from multiple sources, many of which are compilations themselves. I hope you will not disregard them. We may tend to dismiss an occasional inconvenient fact, but we cannot and should not ignore a plethora of facts.

Let's look specifically at a town called Leicester, which abandoned vaccination because of the high number of smallpox mortalities among vaccinated individuals, and instituted their own method to combat the disease. This town illustrates the difference of results between vaccination and the God-given instructions of sanitation and quarantine. Afterward, we will see many accounts from doctors and other educated men who spoke out during or shortly after smallpox epidemics and on through the early 1900s. First, we must establish foundational knowledge regarding smallpox, starting with the difference between inoculation and vaccination.

Today, these terms are usually synonymous, but throughout history mankind has been inoculated with different diseases. A few hundred years ago, people were inoculated with smallpox matter from person to person, which has existed for thousands of years. This inoculation was also called variolation—variola being another name for smallpox. Vaccination was the spread of what was believed to be cowpox—vacca being the Latin word for cow—to impart immunity to smallpox. Before the use of Jenner's inoculation with cowpox, people were inoculated with smallpox (variolous) matter.

"In many cases the operation [inoculation/variolation] was performed by buying a few scabs or pus from someone suffering from natural smallpox and then puncturing the skin with a needle which had been contaminated with the smallpox matter."[22]

Inoculation can mean variolation or vaccination. Some of the following quotes use the terms interchangeably, and some do not; the context usually prevents confusion. Because one could have been inoculated with smallpox, cowpox, or the various "lymphs" that came to be used, inoculation was used synonymously with vaccination even over a century ago.

"Although 2% to 3% of variolated persons died from the disease, became the source of another epidemic, or suffered from diseases (e.g., tuberculosis and syphilis) transmitted by the procedure itself, variolation rapidly gained popularity among both aristocratic and common people in Europe."[23]

Also, Jenner was not the first to suppose cowpox could protect against smallpox. It has been said in science credit goes to the man who convinces the world, not the man to whom the idea first occurs. Jenner convinced the world, but the ideas that led to Jenner's vaccination were brought to England before him. Vaccination evolved, if you will, from variolation, which has superstitious, rather than scientific, roots.

"Mr. Porter, who was an English ambassador at Constantinople in 1755, informs us, (Gentleman's Magazine, for October of that year): 'It is the tradition and opinion of the inhabitants of the country that a certain angel presides over this disease. That it is to bespeak his favor and evince their confidence that the Georgians take a small portion of variolous matter, and, by means of scarification, introduce it between the thumb and forefinger of a sound person. The operation is supposed to never miss its effect. To secure beyond all uncertainty the good will of the angel, they hang up scarlet cloths about the bed, that being the favorite colour of the Celestial inhabitant they wish to propitiate."[24]

Smallpox Vaccination Science: Better Living Through Chimerical Superstition.

This inoculation, or variolation, came to be outlawed in many places.

"After vaccination was introduced, it was ascertained that inoculation added greatly to the number of small-pox cases, and that the mortality was not diminished, but rather increased. Stringent laws were then passed, in different countries, making the practice of inoculation a crime."[25]

While vaccination became compulsory in certain countries.

"The Vaccination Act of 1853 [in England] made vaccination compulsory for all infants in the first three months of life and made defaulting parents liable to a fine or imprisonment. The Act of 1867 extended the compulsory vaccination requirement to age 14, with cumulative penalties for non-compliance."[26]

Imagine having your newborn child's soft skin cut and introducing animal pus for their "health." If you refuse and can't pay the fine, you go to prison. This is how vaccination began. Today, technology has advanced, however, the practice is still just as barbaric and unnatural as it ever was, but these aspects of the process now occur behind the scenes. Diseased animal matter in the human body does not make one healthy, no matter how it is prepared.

"As in the case of other notions which have not commended themselves to popular acceptance, its advocates do not rest content with submitting the matter to rational enquiry, but seek to procure compulsory obedience by law and penalty. This is but a form of violence; and the resort to it is equivalent, therefore, to an acknowledgement on the part of these men, that they are actually in the wrong. Truth can afford to wait the ages; but error would expire on the way."[27]

What you are about to read is much different than what you have been told your entire life. Here is a quick statement to help you understand the accounts to come:

"Un-refutable, scientific evidence proves that smallpox and cowpox are not the same disease, therefore, cowpox vaccination cannot introduce anti-bodies into the system that fight smallpox, as is claimed by the medical doctors."[28]

Doctors also observed that smallpox could be contracted more than once, which would mean the protection provided by vaccination would be very limited for a disease that could repeatedly occur in the same individual. Most importantly, it would behoove us to recognize that God told His people to *separate* from sickness, not spread it; and I will trust God no matter who says the Bible is an "outdated" book full of inaccuracies.

Many will claim that science and Scripture are incompatible because the Bible is inherently unscientific. These statements are also completely untrue. In fact, amazing truths have been recorded in the Scriptures, *thousands of years* before man identified them. A science textbook from 50 years ago would be rife with error, while the Bible is just as accurate today as it was the day it was written.

The Bible tells us about the dividing of the continents (Gen 10:25), the perfect dimensions for a nautical vessel (Gen 6:15), jet streams (Ecc 1:6), and the hydrological cycle (Ecc 1:7). God told us in His Word that there are mountains (Jon 2:5-6) and springs on the ocean floor (Job 38:16), and currents in the oceans (Psa 8:8). The Scriptures also tell us that light travels in a path (Job 38:19-20), air has weight (Job 28:25), man cannot number the stars (Jer 33:22), and that light is divisible (Job 38:24). God's Word explains what we call the First (Gen 2:1-2) and Second (Psa 102:25-26) Laws of Thermodynamics. *These truths were all recorded thousands of years ago in God's Word*. In one of the oldest books of the Bible, God asks Job, *"Canst thou send lightnings, that they may go, and say unto thee Here we are?"* (Job 38:35). Today, we know that radio waves are a form of light, which can indeed send sound. *The Bible is true*, and anything revealed in it will stand the test of time (Isa 46:10). God gave His people the best way to avoid disease, and man's devices designed to accomplish this same task that are in opposition to the wisdom of God's Word are destined to fail.

With these foundational understandings in mind, let's examine the truth about smallpox vaccination based on historical eyewitness accounts.

After nearly 20 years of compulsory vaccination, a deadly smallpox epidemic struck England and many other countries, which brings us to Leicester, where we will see the difference between sanitation and quarantine versus vaccination.

"...in 1872. A severe epidemic of small-pox prevailed in New York, during that year, notwithstanding the fact that the health authorities had claimed the year previous, that the city was thoroughly protected by vaccination."[29]

"What was even more significant, many vaccinated persons in almost every place were attacked by small-pox before any unvaccinated persons took the disease [during the epidemics of 1870-72]. These facts are sufficient to overthrow the entire theory of the protective efficacy of vaccination."[30]

"Despite the actions taken by the government to ensure a very high vaccination rate, a massive smallpox epidemic hit not only Leicester but all of England and other parts of the world in the early 1870s. The epidemic in Leicester resulted in thousands of cases of smallpox and hundreds of deaths, shaking to its core many people's belief in the protective powers of vaccination. ... After the 1872 pandemic, even more people lost confidence in vaccination. They began asking the question as to whether better sanitation, hygiene, improved housing, nutrition, and isolation of cases were the best ways to deal with smallpox."[31]

"**The last decade has witnessed an extraordinary decrease in vaccination, but, nevertheless, the town has enjoyed an almost entire immunity from small-pox**... The method of treatment, in a word, is this: —As soon as small-pox breaks out, the medical man and the householder are compelled under penalty to at once report the outbreak to the Corporation. The small-pox van is at once ordered by telephone to make all arrangements, and thus, within a few hours, the sufferer is safely in the hospital. The family and inmates of the house are placed in quarantine in comfortable quarters, and the house thoroughly disinfected. The result is that in every instance the disease has been promptly and completely stamped out at a paltry expense. Under such a system the corporation have expressed their opinion that vaccination is unnecessary, as they claim to deal with the disease in a more direct and much more efficacious manner. This, and a widespread belief that death and disease have resulted from the operation of vaccination, may be said to be the foundation upon which the existing opposition to the Acts rests."[32]

"The experience of unvaccinated Leicester is an eye-opener to the people and an eye-sore to the pro-vaccinists the world over. Here is a great manufacturing town having a population of nearly a quarter of a million, which has demonstrated by a crucial test of an experience extending over a period of more than a quarter of a century, that **an unvaccinated population has been far less susceptible to small-pox and far less afflicted by that disease since it abandoned vaccination than it was a time when ninety-five per cent of its births were vaccinated and its adult populated well re-vaccinated**."[33]

"How is it that in some of our best vaccinated towns — for example Bombay and Calcutta — small-pox is rife, whilst in some of our worst vaccinated towns, such as Leicester, it is almost unknown?"'[34]

"It must strike the reflective observer as rather singular that all **the recent smallpox outbreaks have made their appearance among populations where the laws enforcing vaccination have been rigorously and systematically carried out**. 96% of births in London are protected by vaccination. May I venture to ask whether medical men who have defended and fostered a system of medical procedure which eighty years' experience has demonstrated a disastrous and humiliating failure ought not to feel honourably bound on public grounds to retrace their steps and confess that vaccination, like other once popular prescriptions of inoculation, bleeding and mecurization, is a serious and mischievous blunder."[35]

"All fallacies classified as science must crumble before investigation. Such has been the fate of all the pretentious theories of earlier medicine, and such is the predestined end of the delusive hypotheses upon which are based many of the medical dogmas of to-day. Of these dogmas I believe the practice known as *vaccination* to be most absurd, and most pernicious. I do not believe that a single person has ever been protected from small-pox by it; while I know that many serious bodily evils and even deaths have resulted from its employment. ... Every physician of experience has met with numerous cases of cutaneous eruptions, erysipelas, and syphilis, which were directly traceable to vaccination, and if these could all be collected and presented in one report, they would form a more terrible picture than the worst that has ever been drawn to portray the horrors of small-pox."[36]

"After a careful study of the history of vaccination and the extensive experience in its use, I am thoroughly convinced that: (1) Vaccination is utterly useless as a preventive against smallpox, that millions of vaccinated persons have died of smallpox. ... (7) Vaccination protects from smallpox only when it kills the patient before smallpox develops. ... (13) Disease as the result of vaccination is the logical harvest from the seed sown. (14) Vaccination has no scientific basis upon which to rest its claims and no analogy in any ascertained principle or law in nature."[37]

"I would not go so far as to say that vaccination has never saved a person from smallpox. It is a matter of record that thousands of the victims of this superstitious rite have been saved from smallpox by the **immunizing potency of death**. But it is a fact that the official statistics of England and Wales show unmistakably that, while vaccination has killed ten times more people than smallpox, there has been a decrease in smallpox concomitant with the decrease in vaccination… It might be appropriately asked, in the words of the **Vaccination Enquirer** (London, Feb. 1947) 'How could an operation that was declining be responsible for the extermination of smallpox?'"[38]

"Records in all countries show that the so-called, **contagious** diseases declined quite rapidly with the introduction of sanitation and nutrition reform, with the exception of **smallpox** and **diphtheria** which were kept active by vaccination in a large number of authenticated cases."[39]

"I have seen too many instances of small-pox in children vaccinated in London, where that process was carried on in the way which the National Vaccine Establishment has recommended as the most efficacious, to retain much faith in its preventive powers, in whatever manner conducted. … But the present epidemic shows too clearly the mortifying fallibility of medical opinions, though founded on the experience of twenty years, and guaranteed by the concurring testimony of all the first physicians and surgeons in the world."[40]

"If this method of disease prevention had ever been successful, smallpox would have been eliminated from the world centuries ago, but smallpox persists in localities where vaccination is compulsory and declines where it is abandoned in favor of sanitation and hygienic measures."[41]

When reading historical eyewitness accounts from the last two centuries, it's amazing that vaccination is credited with eliminating smallpox when that is far from the truth. As the previous quote expressed, if vaccination is what stopped smallpox, why did it take nearly two hundred years for this to happen? Vaccination was very profitable, and protecting the practice has always been of the utmost importance.

"Whoever closely watched the course of the epidemic in Montreal must conclude that vaccination is utterly useless as a protection from small-pox. Much of what transpired in our small-pox hospitals was suppressed, especially whatever was likely to operate against the progress of vaccination, which proves a golden harvest to the vaccinators. But notwithstanding the conspiracy of silence a few official reports pregnant with proof against vaccination, and proving beyond question that a large proportion of the patients admitted into our small-pox hospital had been vaccinated, and that many of them died, some with two and others with three, vaccine marks upon their bodies."[42]

What we have been told is so different from the truth; we must ask, why is this necessary? What is being hidden from us? Why is it so important for everyone to believe that Jenner's vaccination saved humanity? The belief that vaccination eradicated smallpox is the exact opposite of what has been historically recorded. While Edward Jenner was not the first to try to impart immunity by spreading disease, today's medical community certainly follows in his footsteps with inadequate trials and protecting vaccination from receiving negative publicity (which we will see in the next chapter).

"...St. Andrew's, in those days, was not squeamish about granting degrees so long as the fees were forthcoming, Jenner became **Doctor** Jenner for the modest sum of 15 pounds. ... It is true that medical research in the 19th century was not what it is today, but even then the picture of the whole of the College of Physicians and Surgeons swallowing the theory of an unqualified country apothecary, based on one totally unreliable experiment, seems scarcely credible. ...in 1820 he [Jenner] wrote: 'When I found Dr. Woodworth about to publish his pamphlet relative to the eruption (smallpox) cases at the Smallpox Hospital, I entreated him in the strongest terms, both by letter and conversation, not to do a thing that would so disturb the progress of vaccination.'"[43]

And, of course, we hear about the success of Jenner's experiment on eight-year-old James Phipps, but we don't ever hear that James died at the age of 20 from tuberculosis, which many doctors of his day put forth as a side-effect of vaccination. Jenner's son, whom he also vaccinated, was sickly and died of tuberculosis at the age of 21.[44] It is recorded in the second volume of *The Life of Edward Jenner, M.D.* by John Baron that Edward Jenner said of his first vaccination patient:

"Oh! there is poor Phipps...I wish you could see him; he has been very unwell lately, and I am afraid he has got tubercles in the lungs. He was recently inoculated for small-pox, I believe for the twentieth time, and all without effect."[45]

"Vaccination does not stay the spread of disease nor even modify it in those cases who get it after vaccination. It does introduce into the system, and therefore contributes to the spread of, tuberculosis, cancer, and even leprosy. It tends to make more virulent epidemics of smallpox and to make them more extensive."[46]

Jenner was accused of inventing new theories to account for his failed theories. Such as claiming there was something called spurious cowpox—and the lymph produced by this disease did not protect against smallpox—to account for vaccinated individuals contracting smallpox.

Interestingly, we have all been educated to believe that vaccination stopped smallpox, but how many of us have ever seen a modicum of evidence to substantiate this claim? Have you ever seen any rock-solid proof, or have you only heard and read that "vaccines eradicated smallpox," *ad infinitum*? Some statistics favor smallpox vaccination, but many authors, such as James Peebles, Robert Gunn, and Alfred Russel Wallace, have examined these statistics and demonstrated they are not to be trusted. The diagnosing of illnesses in vaccinated individuals as pustular eczema, or some other type of pox instead of smallpox is why statistics support smallpox vaccination as being effective in many cases. Statistics can also be manipulated to be very misleading. In Darrell Huff's book *How to Lie with Statistics*, he tells us: "Like the 'little dash of powder, little pot of paint,' statistics are making many an important fact 'look like what she ain't.' A well-wrapped statistic is better than Hitler's 'big lie'; it misleads, yet it cannot be pinned on you. ... The crooks already know these tricks; honest men must learn them in self-defense."[47]

"There does not exist one single fact, in all the experiments and improvements made in science, which can support the idea of vaccination. The only basis is that of statistics, and these are very unreliable, especially when they are collected for the purpose of sustaining an idea. Yet the enumeration of patients in the smallpox hospitals, will show a larger number vaccinated than unvaccinated. The recurrence of smallpox in the several countries where the law has required everybody to be vaccinated, shows that there is no possible ground for the hope that vaccination can ever terminate its existence on earth."[48]

In his book *Vaccination, a Curse and Menace to Personal Liberty*, Dr. James Peebles tears to shreds the claims and evidence of the vaccinators of his day. He fervently believed in doing what was right, was an ardent supporter of women's suffrage, and helped runaway slaves use the underground railroad to reach safety in Canada. I highly recommend his book if you want to read more about smallpox vaccination.

Over one hundred years ago, it was recognized that the pox some cows had was actually syphilis, which was transmitted from humans as only the milk cows would contract the condition. Even those who believe vaccination stopped smallpox acknowledge that vaccination caused syphilis. *The Lancet* in 1878 stated that those who declared animal lymph could not have "syphilic contamination" were "fallacious."[49] (Another source listed the year of this published statement as 1879.) In 1892, it was also published in *The Lancet* that "there are no means of determining the purity of the lymph or limiting the certainty of its inflammatory effects."[50]

"During more than 60 years the 'vaccination' was carried out from 'arm to arm' with a certain risk of transmission of syphilis."[51]

There was even litigation because of this spread of syphilis.

"Dr. Georg Hübner, the defendant, was accused of having initiated a small epidemic of syphilis by using the lymph of a syphilitic infant to vaccinate 13 infants."[52]

Can you imagine, bearing and raising a child to have him or her given syphilis because of a sanctioned medical practice? Dr. Peebles puts it much better than I ever could:

"May your souls rise up in indignant protest against the sacrilegious invasion of the home which the American constitution has sanctified to liberty; aye, in protest against this infernal rite, becoming the hag of the pit! Think of it, mothers, who would bring your daughters up to be healthy, and clean, and chaste, that your state and city should have delegated the privilege to fee-hunting doctors, to break down the protective [walls] you have builded [sic] about your little ones, and poison the fountains of their lifeblood with a virus of contagious which may mock your solicitude and disappoint the fondest hopes you have cherished for the futures of your posterity."[53]

"By and by came the doctor in his gold-laced cap, with his bottle of 'lymph,' pure from the sores of children or heifer's buttock, and commenced operations."[54]

Vaccinating with whatever lymph du jour available spread much sickness among its unwitting recipients. Some of those who suffered because of "bad lymph" stated, "they would much prefer to have small-pox."[55] One must wonder when this eye-of-newt methodology regarding smallpox vaccination turned into an exact science. Or rather, if it never did, but using a syringe had the advantage of at least appearing more technologically advanced than cutting people and applying animal pus as though it were a healing balm. Vaccination was not much different than its forerunner, inoculation.

"Vaccination...does just what inoculation did—cause the spread of disease."[56]

James Peebles again eloquently and accurately declares:

"The practice of vaccination therefore is utterly opposed to the plain teachings of sanitary science. It is the most untenable dogma in the whole category of medical theories, which has never been demonstrated to be sanctioned by any ascertained law or principle in the healing art. No precious lives have been saved as the outcomes of the vaccine delusion... It has become an added factor for the wider diffusion of cancer, erysipelas, eczema, carbuncles, tumors, leprosy, and last but not least, to relegate the 'bad disease' from its dark, infernal den, domesticate and make it common in the households of the land!"[57]

"It is indeed during seasons of small-pox epidemic that we most fully realize the utter worthlessness of vaccination—if we but use our eyes rationally. Yet with all these facts before them, boards of health in a great number of cities and towns in our own country, dogged on by a motley crew of second-rate doctors, drag forth a mouldy Compulsory Vaccination Act from its pigeon hole, and then order all school children vaccinated, on pain of expulsion from the public school... And then, this vaccination practice has become a business and source of revenue to a privileged class, which neither professional or politician will ever consent that it shall grow less. I say 'professionals' with a qualification, since I have reference only to the mercenary class within the ranks of the medical profession. If this class [does] not constitute a majority, they have 'cheek' sufficient to accomplish a vast amount of mischief. A mercenary doctor, lawyer, or priest is a curse in any community; but when they form cabal and compacts to perpetuate a monstrous practice they augment the curse into a public scourge!"[58]

"If they do not realize the wrong they are perpetuating on the rising generation God pity their professional acumen; and may He with the good angels of heaven especially pity the little children who are turned over to the lance and putrid pus of this modern molock! [sic]"[59]

"But why should the rising generation stand in the way of business? 'Business is business.' 'We must live.' These are corporation and class ethics, without soul, without conscience, cruel as fate, knowing no other object or goal than what self interest dictates. This is the code by which oppressive laws become recorded in our statute books, and are kept there; the code by which otherwise good men will enter into compacts which, in their composite character, become merciless tyrannies—a veritable car of Juggernaut—bearing down its victims without pity and without remorse!"[60]

Many of these quotes regarding smallpox do not speak positively of the doctors that supported compulsory vaccination. I would like to point out that most of these quotes are from men who were doctors themselves, many of whom had seen first-hand the dangers of vaccination, and no protective benefit whatsoever. These negative comments are directed at men who were not worthy of the title physician because they put profit first and lobbied for taking away others' freedom through compulsory vaccination—to increase their income. I do not want to give the impression that I am against doctors in any way; I am against those who conceal the truth and put profit before human life, regardless of their profession.

Throughout the smallpox epidemics, there were always doctors that were true to their profession. Who first did no harm, and put their patient's lives ahead of their pockets. They knew that smallpox was typically a disease of filth, and truly tried to save the lives of those living in the horrid conditions that allowed the epidemics in the first place. I've included a longer excerpt here to show the conditions that afforded not only smallpox epidemics but many other diseases that pose virtually no risk to our children today.

"In March, 1885, my attention was aroused by a report that several cases of small-pox existed in the east end of Montreal. Knowing something of the filthy condition of the certain localities, I made a careful sanitary survey… What I saw I will attempt to describe—what I smelt cannot be described! I found ten thousand seven hundred cesspits reeking with rottenness and unmentionable filth; many of these pest holes had not been emptied for years; the accumulated filth was left to poison the air of the city and make it the seed-bed of the germs of zymotic [infectious] diseases. Further, I found the courts, alleys, and lanes in as bad a condition as they possibly could be—decaying animal and vegetable matter abounded on all sides. Everywhere unsightly and offensive objects met the eyes, and abominable smells proved the existence of disease-engendering matter, which supplied the very condition necessary for the incubation, nourishment and growth of small-pox.

"Knowing well the fearful consequences that would result from the presence of such a mass of filth in such a densely populated part of the city, I gave the widest publicity to the subject, hoping thereby to rouse the municipal authorities to a proper appreciation of the danger that menaced the health of the city. But I was an alarmist; my advice went unheeded and the filth remained as a nest for the nourishment of small-pox, which grew in strength and virulence rapidly, until it swept into untimely graves, from the very localities I have mentioned, thirty-four hundred persons!—victims of municipal neglect. Instead of removing the filth and putting the city in a thoroughly clean, defensive condition by the enforcement of wise sanitary regulations and the adoption of a rigid system of isolation of small-pox patients, the authorities were led by the medical profession to set up the fetish of vaccination and proclaim its protective virtues, through the columns of an ignorant, tyrannical and time-serving press. Day after day the glaring, snaring headlines of 'Vaccinate! Vaccinate!' 'Alarm! Alarm!' appeared in the morning and evening papers. A panic of cowardice and madness followed, and tens of thousands of people were driven (like sheep to the shambles of the butcher), to the vaccinators, who reaped a rich but unholy harvest. ... I did all in my power to convince the authorities and the people of the sad mistake they were making; but ignorance, vaccination and love of money gained the ascendancy, and three thousand four hundred innocents were sent to untimely graves. ...mostly children under twelve years of age, died from small-pox in the very localities I pointed out as abounding in filth; while in the west end...where cleanliness prevailed, there were only a few cases and these sporadics."[61]

The preventative measures of sanitation and cleanliness were ignored, but when smallpox outbreaks began it would be announced throughout the city that everyone should be vaccinated or re-vaccinated. The people living in these wretched conditions were not protected by vaccination, and were more likely to suffer harm or die after having pustulent matter introduced into their already weakened bodies.

"There is nothing in the nature or elements of vaccine pustule or of the ferment of the body of the patient which vaccination creates, to counteract smallpox tendencies, or to eradicate them. It is simply the outcome of a filthy disease. The idea of employing such an agency to purify the constitution from any poison is repugnant to the plainest principles of common sense. The act of vaccination is simply the contaminating of a patient's body with a blood poison. Its most fitting eulogy is that found in the New Testament: Satan casting out Satan. The conclusion is inevitable in each case, it will not [be] done."[62]

Why have you never heard these things? Ignorance and avarice. Most do not know this information; they are convinced vaccines save lives and that smallpox was the first disease eradicated by vaccination. Those who know the truth, yet stay silent, simply love money.

"...in a thorough system of sanitation, which, though diminishing doctor's fees to an alarming extent, gives a most satisfactory result in an enormous reduction of zymotic diseases..."[63]

It is of the utmost importance that trust in vaccination is maintained. (We will soon see even the US government has stated this.)

The only methods that bring what vaccines promise are God's laws of sanitation and quarantine. These, implemented with the abandonment of vaccination, yield the best results. Some of these facts might be hard to accept; research and read these accounts for yourself. Vaccination a few hundred years ago was much different than it is today. Cutting one's skin and administering pus from a diseased animal into a healthy person's body is so contrary to what God said would keep His people from disease; it should not be surprising to the Bible believer that this practice spread sickness.

"Even now, in what is sometimes called **the enlightened age**, we still hear some people, (not so enlightened) claim or repeat what they have been taught to say, that 'vaccination has reduced smallpox and certain other diseases, and has been a blessing to mankind.' Those who make this claim have not bothered to consult authentic records or become acquainted with the most elementary facts on the subject. When traced down, these statements are found to originate from the fabricated propaganda of those who profit from the sale of serums and drugs. Unfortunately many of these people are in places of authority where they influence the thinking of students, writers, public officials, teachers and others who mold the thinking of the masses."[64]

Some of the advertisements to warn the public read as follows:

"Entire Repeal and no Compromise.

Sanitation not Vaccination.

From Horse Grease, Cow-pox, Calf Lymph, and the Local Government Board 'Good Lord Deliver Us.'

Better a felon's cell than a poisoned babe. ...

It is not small-pox you are stamping out, but human creatures' lives."[65]

We have all heard, read, and believed that "vaccination eradicated smallpox," but we never hear about some of the recognized disfigurement and death that were recorded in history. There are horrible pictures of suffering children, completely covered in sores, some with flesh rotting away. There are multitudinous accounts of children dying shortly after vaccination, being sickly for the rest of their short lives after vaccination, and pictures of "poisoned babes" in their small coffins. Entire families were infected with syphilis, and thousands of people, including young children, became lepers for the rest of their lives because of vaccination.

"Even if the cow-pox did afford a certain security against small-pox infection, as Dr. Jenner has represented it, it would still remain a question whether the human race would really be benefited by its universal adoption, since the cutaneous eruptions that have followed have in many instances proved more fulsome than even small-pox itself. That those eruptions do occur after cow-pox infection must be allowed by its most strenuous advocates, being perfectly novel, of a nature unknown before the introduction of vaccination, and peculiar to those who have been vaccinated, and often so inveterate as more than to counterbalance the trivial advantages that we were first led to expect from its introduction. … It must be allowed that the local inflammation excited by the inoculation with this matter, is of a very unfavorable nature, and often ends in a deep sloughing, frequently producing such an adhesion of the muscles of the arm, as very much to confine its motions; and some instances have occurred of the mortification spreading, so as to destroy the life of the child; an instance of which happened in St. George's Fields. The child was inoculated at the Cow-pox Institution, Salisbury Square, Fleet Street; the inflammation of the arm exceeded its usual boundary; on the sixth day mortification ensued, which proved fatal to the child."[66]

"My illustration shows what an ugly blot and what a ghastly risk vaccination is when it can change a healthy child into an object the mother can never look at without a shudder. No consideration in the wide world, save that of its money value, would lead a body of men, claiming some knowledge of pathology, physiology, and chemistry, to retain an observance where such accidents are possible."[67]

"The general declaration of my patients enables me to proclaim that the vaccine notion is not only an illusion, but a curse to humanity."[68]

"I have seen scores of parents who tell me that they honestly believe that their children had died from vaccination."[69]

"I have often—almost daily—heard parents say: 'My children have never been the same since they were vaccinated.'"[70]

"Doctor Schlegel, of Saxony, finds 'the number of unfortunates who suffer from affections of the eyes, ear, speech, and mind infinitely greater now than before vaccination.' … 'The blood of the whole British population is saturated with scrofulous and tubercular diseases, which are more destructive to the youth and flower of the European races than ever were the cholera, plague, or the most bloody wars of Napoleon.'"[71]

"Professor Bock of Leipzic, says: 'I have in a forty years' practice seen far more evil than good from vaccination, and therefore vaccinated my own children only in their third and fourth year, when they were hearty and strong. But I would never have vaccinated them but for an indirect compulsory vaccination.'"[72]

"A sixth case, exceedingly well before vaccination, was never well afterward. Its flesh rotted on [the] slightest scratch of a pin, and now and then broke out in scales and sores; it died when twenty months old. Six other children were vaccinated from this child, not one of whom survived. A seventh case, a healthy baby before vaccination became ever after an indescribable sufferer, and died at nearly eight, his body being literally rotten; father, mother and five other children all remarkably healthy."[73]

"If vaccination had not been persisted in, the great improvements in sanitation and nutrition would have had a better chance to eliminate disease. But as it is, **we now have a greater scourge of killer diseases than at any other time in recorded history**; and much of it is **traceable to vaccination**."[74]

"During the discussion I advised that instead of vaccinating and poisoning the blood of our clean, sweet-faced children, that the doctors, druggists, lawyers, and preachers [all those who performed and profited from vaccination] of San Diego—*all* be vaccinated and the dear, innocent children be spared. This was not a popular presentation to the vaccinationists present, and yet, two-thirds of the ladies cheered me roundly. Oh, that our mothers, wives, sisters, and daughters could vote..."[75]

I hope the many different aspects associated with smallpox vaccination helps you draw the comparison between what happened in the past to what is still happening today. It was a completely different world that allowed the past epidemics of infectious disease, but the same questionable practices and doubts about vaccination are still present.

We see in the past that smallpox vaccination:

—Was not thoroughly tested before used on the public, nor before it was made compulsory.

"It is an undeniable fact that the doctrine of vaccination as a protection against small-pox never underwent a critical scrutiny of the character that would be required in a court of law. Instead, it was assumed upon doubtful and equivocal evidence, and promulgated as proprietary nostrums are to this day thrust upon the notice of the public. It was accepted, as Dr. Creighton aptly remarks, upon terms which will seem incredibly loose to every person who has not already made acquaintance with the standard of logic in the medical profession. Since that, it is taken upon trust, without inquiry, upon the presumption, so often a mistaken one, that a new project, especially if it be a scientific one, had been thoroughly tested and debated on all sides before it received the general assent of its own age."[76]

—Many negative effects are associated with the practice, including a higher likelihood of catching the disease that the vaccine is supposed to protect against. These unsatisfactory results are not disclosed to the public and are largely unknown by most doctors. The unvaccinated who follow God's laws of sanitation are the most well-protected; where man's devices failed, God's succeeded.

"'I do not believe,' says Sir James Y. Simpson, 'that either vaccination or drugs can give absolute security against the inroads of small-pox. When every care has been taken, the vaccinated person has been known to be attacked by the disease. In an epidemic such cases are extremely common.'"[77]

"We have all observed that the various malignant diseases and epidemics leave many persons unscathed. Typhus, typhoid, intermittent fever, Asiatic cholera, attack only those liable from deterioration of physical stamina, worry, undue fatigue, or paralyzing terror. Men and women in a cheerful temper of mind,* self-possessed, in a fair state of health, neat and orderly in their habits, are protected as by a wall of fire. Such are the facts in regard to small-pox."[78]

(*Remember, the Bible says a merry heart does good, *like a medicine*—Pro 17:22.)

"...hygiene is the only preventive agency to be relied upon. The perils of smallpox have been greatly exaggerated. ... Smallpox would never be confluent, except [one] had been in a bad state of health before."[79]

—Known diseases become more prevalent, and new diseases accompany this mixing of diseased animal matter with the human bloodstream. As you can imagine, in the 19th century, it would not have been possible to ascertain every kind of biological material in this putrid substance.

"Consumption follows in the footsteps of vaccination as directly as an effect ever follows a cause. The vaccine poison being the product of decaying animal tissue and often tuberculous in character... 'It is certain...that scrofulous and tubercular diseases have increased since the introduction of cow-pox...' There had also become prevalent since 1806 an entirely new disease, the typhus... Croup had become more common and malignant with children, as well as whooping-cough. ... Dr. Bakewell testified that leprosy had been transmitted... Cancer may also be communicated. The case of Dr. Barnett, of the city of New York, who was infected fatally in 1895 by the accidental inoculation of carcinous matter, shows conclusively that this is possible ...'our generation has gained a far greater susceptibility to the small-pox poison, which will ravage in the above-mentioned diseased forms of the mucous membrane till the feeding of the poison by vaccination, ordered ever by laws, sanctioned by usage, and held up by the Faculty, is forbidden by severe penalty.'"[80]

—Deaths are associated with the practice but doctors and others who profit from vaccination are blinded to the danger, or actively try to keep vaccination from being associated with any negative publicity.

"In a letter of remonstrance he wrote as follows: 'I wish my professional brethren to be slow to publish fatal cases of small-pox after vaccination.' Among our own people in later years this injunction appears to be diligently heeded."[81]

"And yet there are doctors—rather non-studious medical professionals (complimentarily called 'physicians'), who persistently tell the people that vaccination prevents small-pox. They either do, or they do not know better. Do they think—do they observe—do they investigate—do they study and reason? It has been wisely said that he who will not reason is a bigot, he who dare not is coward, and he who can not is an idiot."[82]

—Most did not think, observe, investigate, study, nor reason that there is anything to question regarding vaccination. This is still true today; although many know there is much to question, and will not submit their own children to be vaccinated. A study from 2005 examined the attitudes of parents toward vaccinations, and those most likely to have "no intention to comply with any new vaccination" were defined as: "Determinants of a fully negative attitude were a high education of the parent, being a health care worker…"[83]

This was observed with smallpox vaccination as well.

"Dr. George Gregory, who was himself physician of the Small-Pox Hospital established in London to test and carry out the theories, absolutely refused to permit his own children to be vaccinated."[84]

—Profits are made because of vaccination; the diseases that follow are also treated, and are also profitable.

"Defy the compulsory law, stamp it in the dust, pay fines, be imprisoned if needs be, die the martyr's death—do anything rather than have your children's blood poisoned and their health wrecked for life. … Reader, friend and fellow-citizen, your protection never occupied the smallest corner in the vaccinator's heart. He sows calf-lymph virus to reap a harvest today, and larger, richer harvest in the future."[85]

—Fear is used to manipulate the populace to encourage them to vaccinate.

"The fears of the whole people have been played upon as to the dangers of small-pox, and the sure prevention by vaccination, until nearly the whole civilized world has become physically corrupted by its practice."[86]

—The underreporting of deaths because of vaccination.

"Mr. Alfred Milnes, a statistician who has paid special attention to this subject, expresses the deliberate opinion, that if the officially admitted deaths were multiplied by twelve, it would come much nearer representing the actual number of deaths resulting from vaccination. … 'The facts and figures of the medical profession and of Government officials, in regard to the question of vaccination, must never be accepted without verification."[87]

"And when we consider that these misstatements, and concealments, and denials of injury, have been going on throughout the whole of the century; that penal legislation has been founded on them; that homes of the poor have been broken up; that thousands have been harried by police and magistrates, have been imprisoned and treated in every way as felons; and that, at the rate now officially admitted, a thousand children have been certainly killed by vaccination during the last twenty years, and an unknown but probably much larger number injured for life, we are driven to the conclusion that those responsible for these reckless misstatements and their terrible results have, thoughtlessly and ignorantly, but none the less certainly, been guilty of a crime against liberty, against health, and against humanity, which will before many years have passed, be universally held to be one of the foulest blots on the civilization of the nineteenth century."[88]

—Even though many acknowledged health issues are associated with vaccination, the benefit is believed to outweigh any harm. It is understood that some will be injured because of vaccination, but everyone should be vaccinated for the good of the public, even if individuals do not want to be "protected" this way.

"A dozen or more physicians in these islands, who were practically familiar with leprosy, responded to this circular and gave their emphatic testimony that vaccination was causing a vast amount of leprosy. Their evidence was before the College of Physicians. …in their answer and advice to Lord Kimberly they said: 'The College of Physicians feel they can not press too strongly on your lordship the importance of enforcing the practice of vaccination for the protection of those who are too ignorant to protect themselves.'"[89]

We will see throughout this compilation that all of these things are still happening today in the name of vaccination. The following quote is also from James Peebles, who was a doctor, and is critiquing those who knowingly put profit before the lives of others. Ever since the fall in Genesis, evil men have always existed, and will continue to exist until the end of this present evil world. I use this quote not because I am against the medical profession—quite the opposite! I am against the many evils that are rooted in the love of money. Christians need to be aware of the methods used to keep us uninformed so that we will not submit to paying for things that actually do harm; most doctors are victims of these methods as well.

"A perusal of the history of vaccination is not calculated to excite our veneration toward the medical profession, the older schools of which sanction a species of blood-poisoning with concentrated animal virus in a manner that contravenes the principles of all true science. Their specifics are largely derived from the traditions and superstitions of an ignorant age. All their theories concerning the preventive and mitigating effects of vaccination belong to the category of pseudo-science. The profession knows this to be pseudo-science, and yet with craft and cunning they shun discussion, shelve complaints, evade and mutilate facts, twist statistics, raise false and [irrelevant] issues, make false returns of deaths from vaccinal injuries, dub anti-vaccinators as pestilent agitators, lobby for compulsory vaccination, persecute the true [physician] who restores the sick without medicine, and do many other things which reveal motives foreign to the public welfare. In this domain—the vaccinating branch of the profession—medical practitioners are inversive, reversive, and subversive; they invert the order of nature by creating disease with the pretence of preventing disease; they revert to an ancient superstition which Jenner borrowed from peasant milk-maids, and which Lady Montagu borrowed from the common folk in Turkey; and they subvert the intention of nature by sowing an extra crop of incurable diseases in the name of health—scrofula, cancer erysipelas, leprosy, consumption., etc."[90]

"When the public sees a doctrine which has been exhibited to them as certain, established, and claiming universal acceptance, proved to be faulty in its very foundation, or discovered to be [willful] and despotic in its essential and chief tendencies, many lose their faith in science."[91]

We can't have anyone losing their faith in falsely-so-called "Science," can we?

"Never in the history of medicine has there been produced so false a theory, such fraudulent assumptions, such disastrous and damning results, as have followed the practice of this disgusting rite; it is the ultima thule of learned quackery, and lacks, and has ever lacked, the faintest shadow of a scientific basis. ... The life-blood of nations has become the cesspool of vaccinators, wherein they have poured the foul excretions that are thrown off from diseased beasts, nature adjudging it too vile to contaminate the system of any living creature."[92]

"...for a poorly-informed public have submitted themselves to be made merchandise of by a class of semi-consciousless physicians, until sloth and ignorance will exact a sterner mode of teaching. The object lessons which the plague, with varied zymotic diseases, will furnish, may have the effect of opening the eyes of an apathetic public to the grave violations of physiological law to which they have so long blindly submitted."[93]

"To affirm that there never has been any scientific warranty for a belief in the alleged protectives virtues of vaccination and that its practice is backed by ignorance and indifference, is a sorry charge to make against the medical profession at the close of the nineteenth century; but the charge, I regret to say, is only too true. I know whereof I affirm, for I, too, must plead guilty to the charge. Before discovering my mistake I had vaccinated more than 3,000 victims, ignorantly supposing the disease I was propagating to be a preventative of small-pox. Having taken for granted what my teachers had asserted, I was a staunch believer in the alleged efficacy of vaccination as a prophylactic against small-pox. I remained in this blind and blissful state of ignorance for years, and not until I acquired a little experience in the school of observation and reflection did I discover my faith was pinned to a shameful fraud."[94]

I have given many quotes about smallpox because they are still so relevant today. For some, it is hard to believe error could be accepted as truth by most of the world; but as we can see, regarding vaccination, *this has been going on for over two hundred years.* Deceit, disease, and death are not new aspects of vaccination canon:

They are the foundation.

We will finish with a quote from the Lancet during the smallpox pandemic of 1870-72, and one other quote that is as true today as it was one hundred years ago—but we are trying to prevent other mild diseases rather than smallpox.

"The 'London Lancet' (1871) says, editorially: 'Those who have been building up in their imagination a great and beneficent system of state medicine, under which the great causes of diseases were to be controlled, must abate their hopefulness. It must be admitted that the existing system of public vaccination has been sadly discredited and almost mocked by the experience of the present epidemic.'"[95]

"What an act of insanity it would be to implant the infective products of undefined disease into the bodies of eight thousand healthy children in order to prevent the possible development of a very few mild cases of small-pox! Could absurdity go further than this?"[96]

For many, the foundation of their trust in vaccines has the "eradication of smallpox" as its foundation. I truly hope you will not build any further upon this foundation of sand until you have thoroughly investigated this matter. The smallpox vaccine that had supposedly done so much good, has been done away with.

"England finally banned the vaccine in 1948, despite the fact that it was one of the most widely heralded 'contributions' which that country had made to modern medicine. This action came after many years of compulsory vaccination, during which period those who refused to submit to its dangers were hurried off to jail."[97]

"Although this is not generally known by the public, mass vaccination for smallpox was discontinued in the United States as much because the vaccine caused brain damage and death as because the disease had virtually disappeared."[98]

Polio

What many believe to be the most beneficial vaccine of the last century is one of the vaccines made from an aborted fetal cell line.

"As early as 1936, forced abortions were performed under the US 'Model Eugenic Sterilization Law' of 1922 and used in polio research."[99]

It might be hard to believe that forced sterilization, which would have resulted in forced abortions—unintentional or otherwise—could happen in the United States. This excerpt is from a study titled "Sterilized in the Name of Public Health: Race, Immigration, and Reproductive Control in Modern California."

"In exploring the history of involuntary sterilization in California, I connect the approximately [20,000] operations performed on patients in state institutions between 1909 and 1979 to the federally funded procedures carried out at a Los Angeles County hospital in the early 1970s."[100]

(I believe Stanley Plotkin was asked about abortions which took place due to psychiatric reasons because some of children from these forced abortions were possibly used in his medical research. One of the studies quoted in chapter two noted that the abortion took place for "psychiatric reasons." Abortion is not about choice and never has been.)

Now, here is what you need to know about polio.

"The polio story is a haunting one: long, complicated, and ugly. It's not a story you will have read or that the medical profession will be able to tell. Beyond the smoke and mirrors lie sketchy statistics, renaming of diseases, and vaccine-induced paralytic polio caused by both the Salk, and the Sabin vaccines. Dr. Albert Sabin's oral Polio vaccine continues to cause paralysis in vaccine recipients today."[101]

The "polio story" is not the cut and dry "vaccines stopped polio" that we have all heard. There are many different sides to this story, with many overlapping aspects. I will attempt to put the different pieces of the story into a linear progression, and put the outliers in the most relevant places to get a clear picture of the polio puzzle. To begin, let's compare how poliomyelitis is defined today with how it was defined in the past.

"Myelitis means inflammation of the spinal cord, and polio means grey. Poliomyelitis—or inflammation of the grey matter of the spinal cord—was a term used to describe anyone, usually infants, who had developed paralysis with no apparent cause."[102]

"If a doctor was summoned to the house of a child who had woken up one morning, unable to move, the physician might tell the parents he suspected 'a poliomyelitis' was involved—inflammation of the grey matter of the spinal cord. The term was employed just like we might use the terms *laceration* or *concussion*. Today, when we use the word polio (a shortened version of poliomyelitis), we are referring to a disease—paralysis caused by a specific virus called poliovirus—that came to be associated with inflammation of the spinal cord. A hundred years ago, poliomyelitis (they did not use the shortened 'polio' moniker) would have been freely suggested as a cause whenever someone exhibited paralysis."[103]

"Although most modern scientists of today do not realize it, poliomyelitis was known to have several distinct causes. Laboratory experiments confirmed it. Even if one were to explain the history of the word 'polio' and how poliomyelitis represented the diagnosis of a lesion within the grey matter of the spinal cord, they would be unlikely to accept the broad context in which polio used to inhabit."[104]

To the average person today, polio is specific to poliomyelitis and subsequent paralysis caused by the poliovirus. This inflammation that causes paralysis—poliomyelitis—can be caused by the poliovirus, but many other things can cause inflammation resulting in paralysis. In the past, poliomyelitis was not only caused by the poliovirus. What was once an umbrella term "poliomyelitis," today evokes an association with a specific virus.

"The difference may seem trivial, but it is important to understand the nomenclature from which *polio* descends because, as we will soon see, many things could elicit a poliomyelitis—not just the poliovirus."[105]

"When a modern reader hears the term polio, they inevitably envision paralysis caused by a specific thing—the poliovirus. When it was first discovered, this virus was thought to be the sole microbe responsible for poliomyelitis and was consequently named poliovirus. Although direct experimentation was not typically conducted on humans, it was eventually discovered that the poliovirus could indeed paralyze. It has an affinity for replicating in nervous tissue, and the resulting inflammation can apparently cause the lesions of poliomyelitis. But humans can also be paralyzed by other viruses—not just the poliovirus. ... These facts suggest a reality that will no doubt confuse the modern reader: Poliomyelitis can be caused by many different things. In humans, there are at least several viruses which can cause poliomyelitis. Numerous bacteria in animals can trigger the same results. And to complicate matters, [experiments] created the hallmark signs of poliomyelitis in creatures...poisoned with arsenic. All of them—under the right conditions—could cause examples of poliomyelitis that were indistinguishable from one another, even under a microscope."[106]

For the sake of clarity, we will use the terms poliomyelitis and poliovirus, when the distinction is necessary, instead of using the word polio. (But, when quoting other sources, it will be hard to determine what kind of "polio" is being referenced.)

Poliomyelitis was first called infantile paralysis or teething paralysis. As you can tell by the names, it usually only affected small children, and teething was commonly associated with this paralysis. What does teething have to do with the inability to move? Well, the medicine of those days, as well as commercial teething powders, contained mercury, and mercury poisoning can cause paralysis.

"As medicine, mercury was as ubiquitous as aspirin is today."[107]

"Even Michael Underwood's description of 'Debility of the Lower Extremities' in 1796—considered to be the first modern description of polio—attributes the onset of the disease to 'the instance of teething or of foul bowels,' a pair of events that would have no doubt been treated with mercury-based purgatives."[108]

"Steedman's Teething Powders. Each package contained eight folded up pieces of paper... Each paper contained around 47 milligrams of mercury chloride powder to be placed on the back of the infant's tongue and washed down with milk or water. While they knew the mercury could have toxic effects, the drive to clear the infant's bowels was believed so essential to their health it was thought worth the risk. As such, mercurial medicines were dispensed to infants for nearly any complaint, including the privation of teething. Paralysis is now a known side effect of mercury poisoning, but, at the time, they would not have believed the powder to have this effect."[109]

This was an uncommon disease but became more common through the 19th and early 20th centuries, coinciding with more exposure among the population to mercury and other "medicinal" poisons that caused poliomyelitis, such as arsenic. A pigment called Paris Green also contained arsenic and was used in commercial products. It came to be used as a pesticide and was subsequently ingested. Exposure to arsenic through this compound also caused paralysis. A similar product, lead arsenate, came to be known for "its effectiveness against insects and its harmlessness to the foliage."[110] Lead arsenate was also used as an insecticide, was also ingested, and also caused paralysis. Its use became more prominent and replaced Paris Green as an insecticide. (Paris Green later came to be marketed and sold as rat poison.) A paper was even published as early as 1888 in *The New England Journal of Medicine* titled "On Motor Paralysis and Other Symptoms of Poisoning from Medicinal Doses of Arsenic" (Putnam).

"Arsenic, even in a few hours after its ingestion, may cause distinct lesions of the spinal cord, of the type known as acute central myelitis, or acute poliomyelitis."[111]

"We may have, in the free use of Paris green in the field and garden, one explanation of the frequency occurrence of arsenic in the system."[112]

Coinciding with the widespread use of lead arsenate, and later DDT, as pesticides, the incidence of polio became even more common. DDT was once believed to be completely safe, even to eat.[113] The dangers of these substances were not well known. Because of their toxicity, as their use increased, so did resulting paralysis—especially in the summer when these pesticides were sprayed, and consumers ate the produce covered in these compounds.

"...in the confusion of that time, the ability for the ingested poison to paralyze was lost on all but the most astute physicians."[114]

"No circumstance in the history of poliomyelitis is so baffling as its change, during the last two decades of the nineteenth century, from a sporadic to an epidemic disease."[115]

Another huge factor played a role in this already confusing narrative: injections causing paralysis.

"The injection of any serum, vaccine, or even penicillin, has shown a very marked increase in the incidence of polio, at least 400%. Statistics on this are so conclusive, no one can deny it."[116]

"In Los Angeles (1946) after the largest smallpox vaccination campaign in the history of the city, an epidemic of polio broke out within two weeks after the 'shots' and by the end of the summer there were 26 deaths from this cause and 1,900 reported cases."[117]

"The statistics collected in this inquiry reveal clearly an association between recent injections and paralysis...We must conclude that in the 1949 epidemic of poliomyelitis in this country, cases of paralysis were occurring which were associated with the inoculation procedures carried out within the month preceding the record date of onset of illness…'(Reported in the *British Medical Journal*, July 1, 1950)."[118]

The same study published in the *British Medical Journal* in 1950 also states:

"…there have been reported occasional and sporadic cases of paralysis following the injection of an antigen. This paralysis has sometimes been limited to the limb in which the injection was made; sometimes it has involved other limbs as well. In most cases a diagnosis of poliomyelitis has been made."[119]

Another study, (from 1992) titled "Unnecessary Injections and Paralytic Poliomyelitis in India" tells us:

"Of 262 children with acute polio, 176 had received unnecessary injections <48 h before paralysis and 12 had received diphtheria-pertussis-tetanus or provocative injections."[120]

While it might be hard to accept that injections caused paralysis which would have been diagnosed as poliomyelitis in some instances, here is an aside to clarify some confusion of this facet of the "polio story." As we have established, any paralysis or "inflammation of the brain and spinal cord" would have been considered poliomyelitis. What we now call encephalitis was once called a form of "poliomyelitis," and even today it is acknowledged that, "Rarely polio can present as encephalitis, clinically indistinguishable from other causes of viral encephalitis."[121]

"…vaccination causes polio and its variations such as encephalitis, meningitis, and certain other diseases. …when there is [inflammation] of the brain and spinal cord affecting the nerves and muscles it is called polio."[122]

(These disease classifications have also changed greatly since then.)

"In regions in which there is no organized vaccination of the population, general paralysis is rare. *In patients with general paralysis,* he (Dr. Daraskwiewiez) has never seen smallpox scars, but *smallpox vaccination scars were always present.* … **'It is impossible to deny a connection between vaccination and the encephalitis which follows it.'**"[123]

Today, it is also known that some injections can cause encephalitis and are used to instigate encephalitis in animal experiments.

"...pertussis toxoid is used to induce encephalitis in lab animals. The pertussis vaccine's ability to cause brain damage is thus not only known, but relied upon by clinical researchers studying brain disorders."[124]

"Pertussis toxin (PT) has been widely used to facilitate the induction of experimental autoimmune encephalomyelitis (EAE) in rodents. ... Together, the results suggest that microbial products, such as PT, could contribute to the initiation of autoimmune disease..."[125]

"Usually intravenous or intraperitoneal pertussis toxin is coadministered on the day of encephalitogen administration, and 1 or 2 boosters often are given over the following days."[126]

"Clinical data suggest that PT is a defining and essential toxin for *B. pertussis* pathogenesis and, when formulated into a vaccine, can prevent disease."[127]

As you can see, pertussis toxoid is used to induce encephalitis, and is essential to developing the pertussis vaccine.

"Encephalitis after vaccination was known to produce the same range of disabilities and impairments. 'Atypical' cases could 'mimic' meningitis, viral encephalitis, or poliomyelitis, indeed, according to H.H. Merrit, 'practically all of the acute diseases of the nervous system.'"[128]

Many believe the "...dramatic upsurge of incidence in 1945-1954" of poliomyelitis was caused by "...intensified vaccination with a variety of vaccines leading to provocation poliomyelitis..."[129] The smallpox vaccine could cause encephalitis, but the pertussis vaccine did so much more often, as one can see from this next study published by *Pediatrics* in 1948:

"That vaccination is a cause of encephalomyelitis is well established. ... Either the encephalopathy following pertussis vaccine must be more common, under present conditions of usage, or more severe or more devastating than that following vaccination against smallpox... ...the encephalopathy following pertussis vaccine seems more devastating than the vast majority of the nervous lesions following the use of smallpox vaccine."[130]

While the pertussis vaccine is purported to prevent disease, we will soon see how low its efficacy is. For now, back to poliomyelitis.

Three main causes of paralysis resulted from poliomyelitis: ingestion of medical mercury and arsenic; ingested pesticides that caused paralysis such as lead arsenate and DDT; and vaccine injections. It appears that poor nutrition played a role as well. The lack of protection from proper nutrient intake made these poisons, pesticides, and injections even more likely to cause poliomyelitis. Many nutritionists noted that low calcium, along with other identified deficiencies, were associated with being diagnosed with "poliomyelitis" in the mid-1900s. I'm more apt to blame the presence of aggressive factors rather than the absence of protective ones—I would imagine that low blood calcium's effect in contributing to poliomyelitis pales in comparison to DDT's toxicity. *However* the protective effects of proper nutrition can be quite amazing. There are accounts of diet being used to protect against radiation sickness in Japan at the end of World War II. A proper diet was protective because "Deficiencies are what lead to the uptake of radioactive elements."[131]

"How did they escape? For some time prior to the blast, Dr. Tatsuichiro Akizuki, one of the directors of the hospital, had prepared himself and his coworkers by confining their diet to...miso soup, brown rice, wakame (seaweed), and Hokkaido pumpkin. The strict macrobiotic diet they had lived on apparently protected them."[132]

Dr. Akizuki wrote in his book, *Nagasaki 1945,* that this diet protected him and his coworkers from severe symptoms of radiation sickness which allowed them to continue caring for the sick.

"But my mineral method, I still believe, made it possible for me to remain alive and to go on working as a doctor in Urakami Hospital - despite the fact that I did not possess a very strong constitution and had, moreover, suffered front the effects of the atomic bomb at a distance of about 1,600 metres. ... 'Dr. Akizuki attributes their survival to the drastic diet he forced on everyone in the hospital...'"[133]

That nutrition protected against radiation poisoning is quite significant. Dr. Akizuki described how entire families survived the blast itself, only to die one by one in the following weeks and months from an invisible sickness that they didn't know how to fight. When he realized diet could help protect his patients he did what he could, but many did not have the benefit of being on a protective diet prior to the blast like Dr. Akizuki and the other hospital employees. It was not until after two substantial rainfalls, which washed away the lingering radioactivity, that these deaths subsided. If nutrition can protect in this situation, it can certainly protect from the environmental causes of poliomyelitis as well.

Because there were so many different causes of poliomyelitis, many different nutrient deficiencies correlated the disease, and many different vitamins were considered protective.

"…contact *in vitro* with adrenalin, cortin and vitamin C is equally destructive for both the toxin of diphtheria and the virus of poliomyelitis. …but, so far as development of paralysis is concerned, the constitution of the child is of greater significance than the virus. …in experimental poliomyelitis in monkeys, [researchers] obtained a fair degree of protective effect from [vitamin] C when administered within a rather definite range of dosage. The latter has demonstrated increased susceptibility when vitamin D is withheld in conjunction with a diet otherwise rich in vitamins."[134]

On the island of Nauru in the Southern Pacific Ocean, protection from a polio-like illness was conferred by making dietary changes to ensure adequate vitamin B intake after it had been discovered that infants were dying because of a deficiency.

"The sale of white flour and polished rice was forbidden in multinational stores, whole-meal flour and brown rice being substituted, and the use of whole wheat bread was encouraged. Sweet potatoes were introduced, and the value of eggs and condensed milk was stressed. The sale of sugar was restricted to 1 lb. per adult per week. The use of the yeast-laden cocoanut-sap beverage was again permitted, and the sick babies were treated with an emulsion of the yeast in cod-liver oil. As a result of these measures the infant death rate rapidly fell from 50 to 7 per cent, at which level it has consistently remained. Not only has there been no further reported occurrence of "poliomyelitis" or infantile beriberi, but even leprosy and tuberculosis, which were rife on the island, have been almost completely eradicated."[135]

When looking back at the different dietary and nutritional methods used to treat poliomyelitis, it is not clear what kind of "polio" is being addressed. We will soon read how poliomyelitis was cured using intravenous vitamin C, but the cause of the subjects' "polio" diagnosis is not disclosed. It is not known what was prevented or cured by diet: paralysis specifically caused by the poliovirus, or any number of things which cause poliomyelitis and paralysis. I would guess both because of the possibility that proper nutrition protected against anything that could lead to an incident of poliomyelitis.

Eleanor McBean wrote in her book, *The Poisoned Needle,* that if most doctors "…would care to observe the 'polio season' they would see that it is only at the times when it is *created* by the doctors with their *poisoned needles* and the people with their excessive eating of sugar products and other devitalizing and semi-poisoned foods."[136] Her book also points out that the paralysis that followed vaccination usually "started" or was "confined to the limb in which the injection had been given."[137] She also quotes *Prevention Magazine* from June 1950, which states, "It has been proved that polio can be caused by *cyanide* poisoning."[138] McBean states, "Cyanide is one of the deadly poisons that is permitted to be used in insecticide sprays on farm produce."[139] Her book also contained an excerpt with this statement:

"Nutritionists believe—and rightly so—that, since all polio cases invariably show a lack of blood calcium—and since the eating or large amounts of refined sugar admittedly lowers the calcium in the blood—there must be some connection between the spread of polio and the large amounts of sweets the modern youngster consumes. Soft drinks have been singled out particularly by Naturopaths as the likely culprits for his blood calcium deficiency—which bring on these polio epidemics during the warm months—but candy and other forms of refined sugar, in all probability, will eventually have to share some of the blame. Pregnant mothers, noted for their lack of blood calcium because of the calcium-stealing by the unborn child, are victims of polio 400% more often than other women."[140]

A doctor by the name of Benjamin Sandler successfully and substantially lowered the number of polio diagnoses in his home state of North Carolina by advising a special diet. His book, *Diet Prevents Polio,* can be found online. He advised that fruit be avoided as well as processed sweets. I would not agree with some of his dietary advice, especially avoiding fruits, as the first chapter of Genesis tells us God specifically gave us fruit to eat. But in this case, if the produce would have been covered with dangerous pesticides, avoiding fruit would have been beneficial.

"...after his diet was publicized on the radio and in the newspapers. The incidence of polio was dropped from 2,402 cases in 1948 to 214 cases in 1949 when the country as a whole—39 states—showed an increase in the number of cases from 1948 to 1949!"[141]

With the spraying of pesticides, ingestion of arsenic and DDT covered fruits, summertime consumption of foods devoid of protective vitamins and minerals, perspiration causing loss of minerals, and vaccinations in the summer, a perfect storm would occur resulting in polio outbreaks in the summer months.

"*Prevention*'s editors speculate that one reason why polio might be more prevalent in hot weather is that iodine is lost in the perspiration along with other minerals."[142]

"Amidst discussion around the cause of this new disease, its summertime peak was a constant source of speculation. Perhaps it was a meteorological phenomenon, many surmised. Maybe it was due to flies or mosquitoes that were so prevalent during summer months—perhaps they were carriers of the disease."[143]

There were so many different things contributing to paralysis; even as far back as 1895, a neurologist questioned:

"It is not possible that we group, clinically, under the name of poliomyelitis, several affections which might in strictness be separated; or, in other words, that several different poisons are liable to affect the anterior horns of the spinal cord, exciting results which, though in the main alike, may differ in detail."[144]

Another doctor noted:

"...some of the commonest symptoms seen in our epidemic were entirely foreign to this disease as long described"[145]

Poliovirus, as opposed to poliomyelitis, is a type of enterovirus; enteroviruses are commonly found in the human intestinal tract and usually pose no threat to the nervous system.

"Naturally existing poliovirus was a common bowel inhabitant for millennia, always there, continuously circulating through humans but never causing paralysis until later when something changed. ... Under healthy environmental and dietary conditions, certain populations had antibodies to all three types of the virus and did not develop paralysis or have significant symptoms after infection. ... Isolated native tribes seem to have had no problem with the infections that were plaguing white men, even though blood results showed that the natives were indeed exposed and infected by many of those very same germs. ...

The key question is: What opportunities could have arisen that afforded poliovirus the ability to cause epidemics in the early 20th century?"[146]

"For thousands of years, enteroviruses seemed incapable of causing problems in the central nervous system, but in the late 1800s, enterovirus infection began to be associated with paralysis."[147]

While there were many different mechanisms by which poliomyelitis can occur, what about the genuine cases of paralysis caused by poliovirus? The administration of poliovirus vaccines to prevent the epidemics of poliomyelitis—many of which were not even caused by the poliovirus—accounts for some of these occurrences.

"...it is a disease that occurs only among the most civilized peoples with the highest standards of sanitation, and the like, being unknown among preliterate cultures that have been relatively untouched by civilization. ...polio disappeared in Europe during the 1940s and 1950s without mass vaccination..."[148]

"There have been at least three major polio epidemics in the United States, according to Dr. Christopher Kent. 'One occurred in the teens, another in the late thirties and the most recent in the fifties.' The first two epidemics simply went away like the old epidemics of the plague. Around 1948 the incidence of polio began to soar (Interestingly, this is when pertussis-whooping cough-vaccine appeared, Dr. Kent points out.) It reached a high in 1949, with 43,000 cases, but by 1951 had dropped to below 28,000. In 1952, when a government subsidized study of polio vaccine began, the rate soared to an all-time high of well over 55,000 cases. After the study, the number of cases dropped again and continued to decline as they had in the previous epidemics. 'This time, however, the vaccine took the credit instead of nature.'"[149]

"During the 1962 congressional hearings on H.R. 10541, Dr. Bernard Greenberg...testified that not only did polio increase substantially (50 percent from 1957 to 1958 and 80 percent from 1958 to 1959) *after* the introduction of mass and frequently compulsory immunization programs, but statistics were manipulated and statements made by the Public Health Service to give the opposite impression. ... The 61 percent drop in polio cases in 1954 was credited to the Salk vaccine when it wasn't even in the state!"[150]

"...in 1958 Israel had a major 'type I' polio epidemic *after* mass vaccinations. There was no difference in protection between the vaccinated and the unvaccinated. In 1961, Massachusetts had a 'type III' polio outbreak and 'there were more paralytic cases in the triple vaccinates than in the unvaccinated.'"[151]

Please notice who is being quoted in the following statement.

"...since 1973 more polio cases have been caused by the live virus vaccine as compared to the number caused by the naturally occurring wild virus."[152] Dr. Jonas Salk

"Today, when the man credited with stamping out polio points to the vaccine as the source of the handful of cases which do exist, it's high time to question what we are gaining by using the vaccine on an entire population."[153] Robert Mendolsohn, MD

"Although Salk's vaccine was made ready for use in 1955, the recall and resulting panic meant it would not see widespread use until a couple of years later, when poliomyelitis deaths had already plummeted from 1952's high of 3,145 to 221. Its efficacy was widely doubted by the scientific community and may not even have been capable of preventing poliomyelitis at all. ... The vaccines—despite their admitted inefficacy and late arrival to the scene—were declared a complete success."[154]

"It is quite obvious that vaccination was totally ineffective in preventing paralytic polio [outbreaks]. However, it was concluded that intensified vaccination was the answer."[155]

"The much quoted writer on medical problems, Morris Beale, who for years edited his informative publication, *Capsule News Digest, from Capitol Hill,* offered a standing reward during the years from 1954 to 1960 of $30,000, which he would pay to anyone who could prove that the polio vaccine was not a killer and a fraud. There were no takers."[156]

The poliovirus vaccine does not prevent other causes of poliomyelitis, and in many instances caused paralysis by the poliovirus. With new technology and increased understanding of disease, the many different causes of poliomyelitis were given distinct names.

"Some years ago at the University of New South Wales in Australia, statistics were compiled that showed that the polio vaccine in use at the time had no influence whatsoever on the polio epidemic. Polio comes in cycles anyway, Dr. Burton said, and when it has been 'conquered' by vaccines, and a disease with identical symptoms continues to appear, doctors look for a new virus because they know the old one has been 'wiped out.' 'And the game goes on,' he added. ... Most cases reported prior to July 1, 1958, as non-paralytic poliomyelitis are now reported as viral or aseptic meningitis."[157]

"Before the vaccine was in widespread use, many distinct diseases were naively grouped under the umbrella of 'polio.' Only after the vaccine was widely accepted was there an effort to distinguish poliovirus from other types of paralytic disease. The following list represents a few that could have been categorized and documented as polio prior to 1958. [1] Enteroviruses such as Coxsackie and ECHO [2] Undiagnosed congenital syphilis [3] Arsenic and DDT toxicity [4] Transverse myelitis [5] Guillain-Barre syndrome [6] Provocation of limb paralysis by intramuscular injections of many types including a variety of vaccines [7] Hand, foot, and mouth disease [8] Lead Poisoning

These are all conditions that still exist today and that a polio vaccine could not prevent."[158]

"Many diseases have been renamed and reclassified to conceal the true nature of vaccine-caused disease: instead of paralytic poliomyelitis we have cerebral palsy, smallpox has become monkey pox or white pox."[159]

While it appears there certainly were cases of paralysis caused by the poliovirus that were not caused by poliovirus vaccines, this is normally a virus that does not affect the nervous system. Something about modern living contributes to this enterovirus's entry into the spinal cord. Many scientists, including Albert Sabin, were perplexed as to why this virus seemed to be harmless among native tribes but affected army troops or the wealthy. "It was a pattern that vexes researchers to this day—where disease and death often tracked the abysmal sanitation and nutrition of the poor, polio seemed to affect the rich and well-to-do."[160] One author theorizes that the combination of metals, pesticides, and these viruses in the gut somehow lead to poliomyelitis in general, but also to the poliomyelitis caused by enteroviruses.

"But in a child whose spinal cord was nestled directly against their intestines, mercuric teething powders of lead arsenate might combine with a simultaneous enterovirus infection to create the perfect opportunity for an otherwise innocuous virus to do serious damage. In a gut compromised from chronic metal or pesticide toxicity, many of the enteroviruses that infect millions of people each day may gain the ability to cross over into the nervous system and create the lesions of poliomyelitis."[161]

The "polio story" is more of a nightmare. It is not the story that everyone wants to hear—how man conquered his foe through science, and that many lives are saved thanks to his inventions. The reality is that polio was mostly a man-made problem. It began sporadically with the use of mercury and arsenic, and became more widespread through the commercial use of pesticides. The largest epidemic of polio had these causes, and many cases of paralysis also occurred due to the use of the DPT vaccine, which was made from the pertussis toxoid that is now used to instigate encephalitis. While there were many causes, "polio" can largely be summed up by the widespread use and cessation of pesticides that caused paralysis.

"[Polio] had appeared suddenly in the 1890s alongside the introduction of the pesticide, lead arsenate, and had suddenly disappeared in the early 1950s alongside the abandonment of DDT."[162]

"Infantile paralysis—the affliction that targeted children's legs, arms and sometimes breathing, and so common in the United States between 1894 and 1954—is largely gone, mostly due to the disappearance of pesticides and medicinal metals being administered to children. But it does still occur—often hidden within other diagnoses such as acute flaccid paralysis or transverse myelitis."[163]

Poliomyelitis is a problem in countries where DDT is still used.

"India is currently the world's largest producer of DDT. ... It should come as no surprise that India, despite perpetual polio vaccines, struggles with various forms of infantile paralysis. This problem is so common, a newer label, *Nonpolio Acute Flaccid*

Paralysis (nonpolio AFP), was created to describe those stricken with paralysis not attributable to a poliovirus infection."[164]

"If someone, particularly a child, felt sick during the summer and showed any signs of dysfunction within their central nervous system—paralysis or not—they were going to be diagnosed with a poliovirus infection in the 1950s. ... And as Dr. Greenberg had said at the Chicago conference, this diagnosis would guarantee free care for children courtesy of the National Infantile Paralysis Foundation. As we now know, they most likely did have *poliomyelitis*, but whether it was caused by the infection from the poliovirus specifically was anyone's guess. ...even in 1959, physicians were getting the diagnosis right only 33% of the time. This was not due to ignorance on their part, but a lack of technology combined with suspicion that each disease had a single causative factor."[165]

"In casual discussions of the disease, people will often mention a grandmother or distant relative who had polio. A conversation may come up with someone who, with the distinct limp of a wither leg, may claim they had polio. The truth is—they probably don't actually know what harmed them. They were *told* they had polio, but a definitive answer was nearly impossible for most to have received back then. They most likely had poliomyelitis—inflammation of the grey matter of the spinal cord. Whether a vaccine would have saved them from harm would be complete speculation."[166]

As we've just read, polio does not threaten someone who has a proper diet, a healthy immune system, and is not exposed to poisons that cause paralysis. Many conditions, once diagnosed as polio, are now classified as other diseases which account for much of the "disappearance" of polio. It is now even believed that, "Franklin Delano Roosevelt, the four-term president who directed his sweeping social policies from his wheelchair, may not have been struck by polio but instead by Guillain-Barre syndrome..." which "...occurs after a mild infection, surgery or, rarely, after an immunization."[167] This is ironic, but fits what we have just learned about the polio epidemics of the past.

Ezekiel 47:12 **And by the river upon the bank thereof, on this side and on that side, shall grow all trees for meat, whose leaf shall not fade, neither shall the fruit thereof be consumed: it shall bring forth new fruit according to his months, because their waters they issued out of the sanctuary: and the fruit thereof shall be for meat, and the leaf thereof for medicine.**

This is one of the two times the word "medicine" is mentioned in the Bible. This is a prophetic passage, but the Bible also tells us that herbs are for the *"service of man"* (Psa 104:14), which is applicable throughout human history. Let's look at some information about Dr. Klenner, who used intravenous vitamin C to treat polio. This treatment aligns with what the Bible calls medicine; it is effective, non-toxic, and has been used to cure what modern medicine calls "incurable."

"Klenner made the following remarks: 'It might be interesting to learn how poliomyelitis was treated in Reidsville, N.C., during the 1948 epidemic. In the past seven years, virus infections have been treated and cured in a period of seventy-two hours by the employment of massive frequent injections of ascorbic acid, or vitamin C. I believe that if vitamin C in these massive doses—6,000 to 20,000 in a twenty-four hour period—is given to these patients with poliomyelitis none will be paralyzed and there will be no further maiming or epidemics of poliomyelitis.' ... Although Klenner managed to publish his landmark article only a month later, which documented his cure of 60 out of 60 cases of polio during the 1948 polio epidemic, his comments at the Annual Session apparently were little heeded and quickly forgotten. ... It is also very significant to note that *none* of the 60 patients treated by Klenner had any of the residual deformities so characteristic of many polio survivors."[168]

The following quote is from a different section of the previously cited book, *Curing the Incurable: Vitamin C, Infectious Diseases, and Toxins,* because of the change in disease classifications, but Dr. Klenner successfully treated other diseases that once could have been diagnosed as "polio."

"...Klenner repeatedly cured patients who were already comatose with encephalitis. The dramatic response of viral encephalitis to vitamin C is especially impressive when one considers how many medicines cannot penetrate into the nervous system either partially or totally."[169]

Imagine that—what God calls medicine is effective in treating many of man's ailments.

Measles and Whooping Cough

The slight increases in measles and whooping cough are labeled and broadcast by the media as "outbreaks," which are subsequently blamed on those who choose not to vaccinate their children; the scientific literature reveals something quite different. MMR is also one of the vaccines made from an aborted fetal cell line (because of the rubella component). It is important to mention that, in industrialized countries, under proper sanitation practices with proper nutrient intake, these infectious diseases simply are not the scourges they once were. There was even a pediatric clinic that "…used to induce measles in children with serious renal diseases (nephrotic syndrome) as means of healing or at least substantially improving the condition. Auto-immune disease like asthma, lupus erythematosus or [eczema] also either disappeared or greatly improved after the child contracted and overcame measles."[170]

"First in 1935, I was in private practice in the coal-mining town of Bedlington, England when the triennial measles epidemic struck. Walking or cycling, I would visit the homes of sick children, and in one day would see 20-30 new measles cases. Those were the Depression years, and the children's diets were decidedly subnormal. It was also the days before antibiotics, so that the treatment was mostly symptomatic and ineffectual. Yet out of more than 500 sick children under my care, not a single one died."[171]

Also, the medical literature indicates a significant problem with this vaccine's efficacy.

"While studying thousands of pages written on vaccines I have not found a single paper which would demonstrate that in epidemic situations only unvaccinated children contracted the diseases. Even during vaccine trials many children contracted the diseases against which they were vaccinated… Not only diseases like whooping cough can affect seriously both vaccinated and unvaccinated children (based on hospital admissions), but there is a new disease - atypical measles - which is an especially vicious form of measles only affecting vaccinated children and with a considerable mortality rate."[172]

"Many studies show that the measles vaccine isn't completely effective and that a significant proportion of those infected in measles outbreaks (>60%) had been vaccinated. … Vitamin A deficiency is a commonly observed symptom of measles. The severity of complications have been linked to the degree of Vitamin A deficiency…"[173]

"The outbreaks among preschool-age children indicate deficiencies in the implementation of the national measles-elimination strategy. However, the extent of measles transmission among highly vaccinated school-age populations suggests that additional strategies, such as selective or mass revaccination, may be necessary to prevent such outbreaks."[174]

The next time a measles "outbreak" is hyped by the media (who fail to mention that these outbreaks occur in highly vaccinated populations), please remember this study:

"Measles is the most transmissible disease known to man. During the 1980s, the number of measles cases in the United States rose dramatically. Surprisingly, 20% to 40% of these cases occurred in persons who had been appropriately immunized against measles. In response, the United States adopted a two-dose universal measles immunization program. We critically examine the effect of vaccine failure in measles occurring in immunized persons. ... We found 18 reports of measles outbreaks in very highly immunized school populations where 71% to 99.8% of students were immunized against measles. Despite these high rates of immunization, 30% to 100% (mean, 77%) of all measles cases in these outbreaks occurred in previously immunized students. In our hypothetical school model, after more than 95% of schoolchildren are immunized against measles, the majority of measles cases occur in appropriately immunized children. ... The apparent paradox is that as measles immunization rates rise to high levels in a population, measles becomes a disease of immunized persons[!]. Because of the failure rate of the vaccine and the unique transmissibility of the measles virus, the currently available measles vaccine, used in a single-dose strategy, is unlikely to completely eliminate measles. The long-term success of a two-dose strategy to eliminate measles remains to be determined."[175]

Immunized: "You keep using that word, I do not think it means what you think it means." Pop culture jokes aside, although immunization is used interchangeably with vaccination, it should not be: vaccination does not provide true immunity. (But for our purposes, the common understanding of immunization is used.)

The media's use of fear to promote vaccination is not a new method:

"...and well do these gentry understand how to multiply their fees. A few cases of small-pox are reported, and immediately a rumor is started which is taken up by the press, and a small-pox scare is soon spreading terror among the populace. Then the vaccination harvest follows."[176]

While "vaccine hesitancy" is labeled an upcoming health crisis and is blamed for some illnesses' resurgence, the failure of these vaccines to perform as promised is never mentioned.

Another inconvenient fact: vaccination actually weakens the immune system.

"…they [vaccines] accomplish what the entire immune system seems to have evolved [or rather, was designed] to prevent. They place the virus directly into the blood and give it access to the major immune organs and tissues without any obvious way of getting rid of it. These attenuated viruses and virus elements persist in the blood for a long time, perhaps permanently. This in turn implies a systematic weakening of the ability to mount an effective response not only to childhood diseases but to other acute infections as well. … The ability to mount a vigorous, acute response to infectious organisms is a fundamental requirement of health and well-being."[177]

"Moreover, according to the World Health Organization, the chances are about 15 times greater that measles will be contracted by those vaccinated for them than by those who are not."[178]

It is foolish to weaken the *entire* immune system to *possibly* be *temporarily* "immune" to a disease that isn't deadly—especially when mild childhood illnesses actually benefit the immune system, leading to a decreased cancer risk later in life.

"Exposures to febrile infectious childhood diseases were associated with subsequently reduced risks for melanoma, ovary, and multiple cancers combined, significant in the latter two groups. Epidemiological studies on common acute infections in adults and subsequent cancer development found these infections to be associated with reduced risks for meningioma, glioma, melanoma and multiple cancers combined, significantly for the latter three groups. Overall, risk reduction increased with the frequency of infections, with febrile infections affording the greatest protection. In contrast to acute infections, chronic infections can be viewed as resulting from a failed immune response and an increasing number have been associated with an elevated cancer risk."[179]

"The present study was designed to investigate the hypothesis that febrile infectious childhood diseases (FICDs) are associated with a lower cancer risk in adulthood… The study consistently revealed a lower cancer risk for patients with a history of FICD. The strongest associations were found between patients with non-breast cancers and rubella respectively chickenpox. A strong association was also found with the overall number of FICD both 'classical' (measles, mumps, rubella, pertussis, scarlet-fever and chickenpox) and 'other.'"[180]

Dr. Robert Mendelsohn appropriately asked: "Have we traded mumps and measles for cancer and leukemia?"[181]

The same Dr. Klenner, who treated polio with intravenous vitamin C, also treated his own daughters (what men do with their own children speaks much louder than anything they say) with oral vitamin C. Low doses caused the symptoms to clear, and the treatment was discontinued when it appeared the infection was gone. Notice what happened when the treatment was ended too soon, and the infection returned (it appears in trying to spare his daughters from the pain of injecting vitamin C, he actually prolonged their illness):

"Klenner was able to demonstrate that the measles could be symptomatically controlled but not eradicated by this pattern of vitamin C dosing for a period of 30 days. At this point he gave 1,000 mg every two hours around the clock for four days, and the infections were permanently eradicated. Klenner noted that for 'the first time a virus infection could be handled as if it were a dog on a leash.' This single clinical experiment conducted by Klenner probably best exemplifies that vitamin C simply must be given in high enough doses for a long enough period of time to effectively kill an invading virus. Following the successful treatment of measles in his daughters, Klenner proceeded to treat new cases of measles with intravenous or intramuscular vitamin C. He noted that he could achieve complete control of the disease within 24 to 36 hours after starting the treatment... Klenner also noted that his patients were able to develop complete immunity to recontracting measles even when he had intervened early and cured the disease before the rash was ever allowed to develop."[182]

Now, we will look at studies which demonstrate the low efficacy of the pertussis vaccine. It would appear the whole cell DPT can be "relied upon" to cause encephalitis more than it can be relied upon to prevent whooping cough.

"Although claimed to be 88 per cent effective among children of 7-18 months, during a nationwide epidemic of whooping cough in 1993, a group of researchers discovered that 82 per cent had completed their full complement of DPT vaccines. Others have commented that the whooping cough vaccine is only to be 36% effective. ... There is evidence of declining vaccine immunity illustrated by transmission of mumps, measles, rubella, polio, Hib, Hepatitis B, smallpox, diphtheria, varicella, whooping cough, etc."[183]

"Since the decline of pertussis immunisation, hospital admission and death rates from whooping cough have fallen unexpectedly."[184]

In one book, the last study was quoted and followed by this statement:

"The same was experienced in Sweden after vaccination against whooping cough was discontinued in 1979."[185]

"Among the best evidence of the dangers of pertussis vaccine is the paper presented by Hutchins *et al*. (1988). The incidence and mortality from whooping cough fell steadily between 1922 and 1975. In 1978 a nationwide immunisation initiative was begun; legislation was passed requiring proof of immunisation for school entry at five to six years of age. Almost immediately the overall incidence of pertussis in the United States trebled, especially in the age group below six months. This age group also experienced the highest mortality from whooping cough."[186]

The Hutchins et al study states:

"As a result of immunization laws, vaccine coverage levels against pertussis at school entry have been greater than 95% since 1980. ... During the period 1980-1986, a total of 17,396 cases of pertussis was reported to CDC by weekly telephone reports. The annual incidence of reported pertussis rose during this period from 0.5 cases per 100,000 population to 1.7/100,000. Infants less than 12 months of age had the highest average annual incidence, estimated at 32 cases per 100,000. ... Rates of hospitalization and complications such as pneumonia, seizures, and encephalopathy associated with pertussis were highest in children less than 6 months of age and declined progressively with increasing age."[187]

One study questioned, "...why is whooping cough increasing in the US despite record rates of pertussis vaccine uptake...?"[188] This study suggests: "The startling global resurgence of pertussis, or whooping cough, in recent years can largely be attributed to the immunological failures of acellular vaccines..."[189] The study's main author made this statement:

"This disease is back because we didn't really understand how our immune defenses against whooping cough worked, and did not understand how the vaccines needed to work to prevent it... Instead we layered assumptions upon assumptions, and now find ourselves in the uncomfortable position of admitting that we may [have] made some crucial errors. This is definitely not where we thought we'd be in 2017."[190] Christopher J. Gill

"Vaccination is not a simple, straight-forward cut-and-dry issue. It is complicated. The diseases are complicated and, moreover, the immune system is very superficially understood by even the most accomplished immunologists today. ...'the immune system remains a black box,' says Garry Fathman, MD, a professor of immunology and rheumatology and associate director of the Institute for Immunology, Transplantation and Infection...'Right now we're still doing the same tests I did when I was a medical student in the late 1960s...' It's staggering complex, comprising at least 15 different interacting cell types that spew dozens of different molecules into the blood to communicate with one another and to do battle. Within each of those cells sit tens of thousands of genes whose activity can be altered by age, exercise, infection, vaccination status, diets, stress, you name it. ... That's an awful lot of moving parts. And we don't really know what the vast majority of them do, or should be doing... **We can't even be sure how to tell when the immune system's not working right, let alone why not, because we don't have good metrics of what a healthy human immune system looks like. Despite billions spent on immune stimulants in**

supermarkets and drugstores last year, we don't know what—if anything—those really do, or what 'immune stimulant' even means."[191]

"Nobody—not even the most educated immunologists—understand or can describe the complete cascade of events that occurs after injecting a vaccine. If physicians realized how little is known today about the immune system and vaccines, they would be duty bound to tell patients that there are no accurate scientific answers."[192]

"No child is born with an intact immune system. During the first 2 years of life the immune system is trying desperately to organize. Despite extravagant claims by the medical community, how that immune system is actually assembled by the body is still largely unknown. What is known for certain is that subjecting the infant's immune environment to an array of manmade pathogens, preservatives, and adjuvants absolutely can have a detrimental effect on the formation of that child's brain and nervous system. All scientists - including the vaccine manufacturers - admit as much."[193]

Vaccines are considered "effective" when they produce antibodies, but this is not a true measure of immunity.

"The clinical evidence for vaccines is their ability to stimulate antibody production in the recipient. What is not clear, however, is whether or not antibody production constitutes immunity. For example, agamma globulin-anemic children are incapable of producing antibodies, yet they recover from infectious diseases almost as quickly as other children. Furthermore, a study published by the British Medical Council in 1950 during a diphtheria epidemic concluded that there was no relationship between antibody count and disease incidence; researchers found resistant people with extremely low antibody counts and sick people with high counts. Natural immunization is a complex interactive process involving many bodily organs and systems; it cannot be replicated merely by the artificial stimulation of antibodies."[194]

"Any toxic or poisonous stimulus may provoke some sort of immune reaction - some kind of inflammation, or fluxing out. So what? That is wholly separate from adding to the body's cumulative memory of how to defend against specific disease processes. ... When the body acquires such memory on its own, with no drugs, we have **natural immunity**. The body now has a new, permanent weapon to protect itself from specific unfriendlies in the environment."[195]

"The whole process of creating a vaccine is to make some disease agent gradually weaker and weaker, until it is below the threshold of causing a reaction in 99.9% of children. That is, it's too weak for the body to recognize as foreign, and can penetrate far deeper into the body than nature would ever have allowed. The immune system DOES NOT get triggered the way it would by a foreign protein. We've tricked the body into **not** responding. Such a process is outside the normal course of biological events; nothing like this ever occurs in nature. By contrast, the natural disease itself is strong enough to trigger the whole chain of normal immune responses. Immunity cannot be forced upon the body. The body must recognize the challenge of an actual foreign invader, and then be given the opportunity to put the pieces of the puzzle together all by itself, to figure out what combination of its own immune powers can recognize, surround, and overcome the foreign agent."[196]

True "immunity cannot be forced upon the body." Man cannot do what God does.

Romans 11:33 **O the depth of the riches both of the wisdom and knowledge of God! how unsearchable are his judgments, and his ways past finding out!**

Psalm 119:73 **Thy hands have made me and fashioned me: give me understanding, that I may learn thy commandments.**

Why is man's knowledge so inept when it comes to the immune system? Because God said that *He* heals, and that *His ways are past finding out* (Deu 32:39; Rom 11:33; See also Ecc 8:17 and 11:5). God, who made our immune system, tells us to be clean and separate from sickness. Man, who can't fully understand the immune system, and usually benefits financially from vaccines, says we need them to maintain health. I trust God and His Son, who said, "*...they that are whole need not a physician...*" (Luk 5:31—Notice that this was penned *by a physician,* Col 4:14). While this statement had a deeper spiritual meaning, that meaning was based on a *physical reality*. Healthy individuals, especially babies and small children with *developing* immune systems, do not need any aborted fetal matter injected into their bodies to maintain health. Jesus created everything (Jhn 1:3), He has superior knowledge of physical matters. Jesus did not make this statement flippantly; Jesus never said anything by mistake. Jesus loves children and does not advocate harming them; you can trust His words.

Matthew 17:5 **While he yet spake, behold, a bright cloud overshadowed them: and behold a voice out of the cloud, which said, This is my beloved Son, in whom I am well pleased; <u>hear ye him</u>.**

John 3:12 **If I have told you earthly things, and ye believe not, how shall ye believe, if I tell you of heavenly things?**

It takes greater faith to believe what Jesus said of heavenly things than it does to believe what He said of earthly things; do you believe Jesus' words, or do you trust man? One of the best ways to show the superiority of Scripture over man's knowledge is to look at the practice of bloodletting in the semi-recent past. For hundreds, if not thousands, of years bloodletting was one of the most common "medical" procedures performed. This harmful practice weakened patients and promoted infection. For thousands of years, the Bible has stated, "*...the life of the flesh is in the blood...*" (Lev 17:11). One phrase from God's word held more truth than every school of medicine in the world. Man's wisdom was wrong, and God's was right. Moreover, this statement was true literally *and* figuratively. Blood sustains life, and examining blood for biomarkers gives a snapshot of the health (life) of the individual from which it came. It was deemed foolish to believe the Bible was right over the "medicine" of the day (as it is today), yet that one statement was true, and everyone who said otherwise was *wrong*.

The same school of medicine that performed bloodletting also once treated fevers by withholding water, light, and fresh air from the sufferer.

"The light of heaven and the free air were excluded; pure cold water was 'sure death!' The life-blood was drained off through the puncture of the lancet; the mouth and throat and stomach were corroded with mineral poisons—and all this was part and parcel of the 'healing art' of those days. From the time of Jesus Christ to the present it has been the same. The 'woman which had the issue of blood for twelve years, and had suffered many things of many physicians, and had spent all that she had, and was nothing better but rather grew worse' it seems had much the same experience with the doctors which each generation from that time to the present has repeated."[197]

Who knows how many were killed by the medical practices of those days. Just because something is practiced and promoted by a large section of the medical community, or the entire world, does not mean it is best. *Everything* **must** be compared to God's word. Imagine parents not letting their fevered children have water—doctor's orders. Today, we do something even more unnatural for the sake of our children's "health": aborted fetal cell lines. Injecting aborted fetal DNA into our children might have the veneer of being a sophisticated method of disease prevention, but at its core is even more nonsensical and barbaric than bloodletting.

The practice of bloodletting was promoted by most of the medical community. Notice the potential for profit as this is something that can be done to anyone for a fee, repeatedly. But this practice was in opposition to what the Bible said. The Bible also tells us in Genesis 9:4, Leviticus 17:14, Deuteronomy 12:23 that the life of the flesh is in the blood. These statements contradicted the medical community, but the Bible's statements were true while the contradicting claim of the world was not. Jesus' statement from Luke 5:31 is also true; it is true literally and figuratively, just like the verses about blood being the life of the flesh. His word is to be trusted over anything this world says that is contradictory. Man has discovered it is much more profitable to make every healthy person believe they need to pay for something to keep them well than to treat the much smaller group that is actually sick. (Paraphrased from Dr. Walter Hadwen.[198])

A doctor who looks at a perfectly healthy baby and says that they need vaccines to keep them healthy is contradicting the words of Jesus, and *is wrong* (Luk 5:31). Furthermore, vaccination has not been demonstrated to be nearly as effective nor safe as it is purported to be. When we eat healthy food (the way God made it) and follow His laws of sanitation, infectious diseases are uncommon, with mortalities being even more rare.

"...nutrition status is what drives a child's ability to fight infections. Low nutrition potentiates susceptibility to infection, and nutrition status declines during infection. Ample and strong data agree—data that show that one-third to one-half of the millions of deaths in children across the world, caused by measles, diarrheal diseases, pneumonia, AIDS, and malaria, are due to undernutrition, not undervaccination. Data show that even mild malnutrition changes how often kids get sick and whether they live or die from an infectious disease. ... Now it seems that the data on child nutrition are ignored—parents are told that the only way to have healthy children is to fully vaccinate them. This is a myth that the body of evidence in child nutrition does not support. What *is* well supported is that nutrition drives learning, growth, development, and immune function in children."[199]

Also, when we use technology to develop medicine that aligns with what *God's word specifically calls medicine,* the deaths, suffering, and discomfort from these diseases are greatly reduced.[200]

Even secular authors trying to warn the public about the dangers of vaccination break down this issue into two opposing arguments: "...either the body can learn its own immunity, or else it is largely inadequate, and requires help from the genius of medical science for survival."[201] This same book also states:

"Here's what vaccines really are: **manmade mutation of pathogens, cultures in genetic material of both human and animal origin, which are injected directly into the developing immune systems of infants and children.** ... The Well Baby Program introduces a human being to a lifetime of dependency on organized medicine. If a child can grow up healthy without doctors and drugs, this posits the dangerous concepts of self-reliance and trust in the body's own natural healing [abilities]."[202]

Interestingly, some who believe in evolution understand and appreciate God's designs (rather than the Designer) more than God's own followers. God designed the body to work well; it does not heal solely of itself like the last comment suggests, but follows God's command to do so. God also gave us additional wisdom in His word to keep us well. Rules of cleanliness and separation from sickness were given to the Jews by God; they have been recorded in the Bible for a few thousand years. Following God's laws caused infectious diseases to nearly disappear, and man claims this was because of vaccines. Man takes credit for the things freely given of God, and gets very rich in the process.

The God-given directions to avoid disease are simple and effective; man cannot do anything better than God. Are we to believe God wanted man to be completely powerless against infectious disease until 70 years ago, when vaccine use became widespread? I would say this does not align with what the Bible tells us about God.

Psalm 145:9 **The LORD is good to all: and his tender mercies are over all his works.**

Even the modern advent of antibiotics was not an invention of man, but rather a discovery of something God placed in nature (see Job 12:8). It was observed that certain molds had antibiotic qualities; this led to the discovery, development, and widespread use of antibiotics. Thanks to antibiotics, untold numbers of lives, and limbs, have been saved.

"The mold which came to be known as 'Fleming's Mold' had been identified as *Penicillium notatum*. This organism was originally isolated from a decaying hyssop plant. In the *Bible*, Psalm 51:7, is the sentence, 'Purge me with Hyssop and I shall be clean.' Can we make a case for this as the discovery of penicillin?"[203]

God is good to all, and His tender mercies are over all his works; they give life as do God's laws. While we are not under the law of the Jews, God gave these laws for a reason, and they are recorded in Scripture "*...for our learning...*" (Rom 15:4). After all, we do not wash our hands because of the law, we wash our hands because *it is wise.*

The Bible constantly affirms that God's words and laws are not only what keep us safe, but keep us alive. Proverbs 7:2 says: *"Keep my commandments, and live; and my law as the apple of thine eye."* David wrote in Psalm 40:11 that the LORD's lovingkindness and truth preserve him. Psalm 119, the longest chapter in the Bible, is all about the importance of following God's laws because they lead to life. Please understand this emphasis on the law is used to show that disobeying God's command not to kill by using aborted fetal cell line vaccines will never lead to health; this is not to say New Testament Christians have any claim to the promises made to Israel.

Psalm 119:77 **Let thy tender mercies come unto me, that I may live: for thy law is my delight.**

Proverbs 12:10b ...**but the tender mercies of the wicked are cruel.**

Man claims that the best way to avoid infectious disease is through vaccines, some of which are made from cell lines from the bodies of aborted children. The mercies of the wicked are cruel, indeed. But as we have seen, *this claim is not true*. Why have innumerable babies been killed for medical experiments to make cell lines for a practice that is not responsible for the dwindling of infectious diseases? The end does not justify the means **if the means are not even responsible for the end**.

A few hundred years ago, choosing to believe Leviticus 17:11 over the inertia of assumed medical dogma of bloodletting looked foolish. God chooses to use the foolish things of this world to confound the wise (1Co 1:27). Furthermore, it pleases God to save man through means that this world esteems foolish (1Co 1:21). Many consider it the epitome of stupidity to refuse these vaccines. Yet, the vaccinated experience disease much more often than the unvaccinated, as we will see next.

Psalm 119:51 **The proud have had me greatly in derision: yet have I not declined from thy law.**

Psalm 146:3 **Put not your trust in princes, nor in the son of man, in whom there is no help.**

1 Corinthians 1:18 **Let no man deceive himself. If any man among you seemeth to be wise in this world, let him become a fool, that he may be wise.**

Part II

2 Peter 1:5 **And beside this, giving all diligence, add to your faith virtue; and to virtue knowledge;**

Psalm 119:66 **Teach me good judgment and knowledge: for I have believed thy commandments.**

"O, God, since Thou hast laid the little children into our arms in utter helplessness, with no protection save our love, we pray that the sweet appeal of their baby hands may not be in vain. Let no innocent life in our city be quenched again in useless pain through our ignorance and sin. May we who are mothers or fathers seek eagerly to join wisdom to our love, lest love itself be deadly when unguided by knowledge."

"Anna's Wail Made John a Prize Baby,"

The New York Times, 7 May 1916

Chapter 4

Many Possible Doubts

Jeremiah 7:5 **For if ye throughly amend your ways and your doings; if ye throughly execute judgment between a man and his neighbour;** 6 **If ye oppress not the stranger, the fatherless, and the widow, and shed not innocent blood in this place, neither walk after other gods to your hurt:** 7 **Then will I cause you to dwell in this place, in the land that I gave to your fathers, for ever and ever.** 8 **Behold, ye trust in lying words, that cannot profit.**

John 5:14 **Afterward Jesus findeth him in the temple, and said unto him, Behold, thou art made whole: sin no more, lest a worse thing come unto thee.**

"The theory of vaccination states that by giving a person a mild form of a disease via the use of immunizing agents, specific antibodies are produced that will protect the organism when the real thing comes along. 'This sounds simple and plausible enough except that it doesn't quite work that way.'"[1]

"The greatest threat of childhood diseases lies in the dangerous and ineffectual efforts made to prevent them through mass immunization."[2]

"The fact that vaccination is either the (a) adding of a disease to a body already weighted with disease, or else, (b) the giving of a disease to a perfectly healthy body, is the medical reason why some of us (medical doctors) oppose vaccination and urge its abolition. A perfectly healthy body is proof against natural disease; a diseased or unhealthy body is handicapped by being given another (inoculated) disease. **A diseased or unhealthy body can never be rendered healthy by inoculation or vaccination.**"[3]

"While the myriad short-term hazards of most immunizations are known (but rarely explained), no one knows the long-term consequences of injecting foreign proteins into the body of your child. Even more shocking is the fact that no one is making any structured effort to find out."[4]

Now, we must also ask, *can* it be beneficial to inject ourselves with the DNA of murdered children? Would God ever *allow* this to bring health? Is this not more likely to be harmful to our health? Would this practice not have intrinsic negative health effects, as do other sins that greatly harm our bodies? Everyone has heard the mantra, "vaccines are safe and effective," what we don't hear are the many, many studies that show the negative health effects associated with vaccination.

This section quotes a lot of studies; bear with me, we will get back to the Bible. I'm making the resource *I wish I'd had* when I tried to navigate through all of this confounding vaccine information. I'm operating under the assumption you are a diligent parent who wants to invest a small amount of time to learn what is best for your child. The following excerpts might be a little tedious, but I have saved you time by reading and consolidating them. When I researched for the sake of my own child, I knew that I was hungry for the truth, and wanted as much relevant information as possible. When I say there are many possible doubts—that's an extreme understatement.

"...any possible doubts, whether or not well founded, about the safety of the vaccine cannot be allowed to exist in view of the need to assure that the vaccine will continue to be used to the maximum extent consistent with the nation's public health objectives."[5]

Stated by the FDA in the
Federal Register, 1984

This statement was made specifically about the polio vaccine; it is logical to reason this policy extends to all vaccines. Doubts about vaccine safety "cannot be allowed to exist." This is justified by the belief that if vaccine rates are lowered, diseases will return, and many more deaths would occur. It is understood susceptible children will be maimed and killed because of vaccines; which is why there is a government Vaccine Adverse Event Reporting System (VAERS), and a small amount of money from every vaccine is set aside to compensate vaccine-injured families. Supposedly, this number of deaths is less than what would be caused by the return of infectious diseases.

"Vaccination is a serious medical intervention for healthy individuals against a disease that *might* occur. The decision whether to vaccinate has important, potentially life-altering ramifications. While vaccines protect some, they unquestionably harm and cause death to others. For government to compel a potentially life-threatening product sets vaccination apart from other medical procedures. Public health officials tell us that those who suffer adverse reactions are exceedingly few. The basis for such an optimistic claim is doubtful. ... When we speak of the greater good, we invoke a higher moral authority claiming that we will benefit more people and save more lives. Nevertheless, there is an uncomfortable underside to the argument. Can we sacrifice one to save ten? Five to save six?"[6]

Good will never come after deviating from God's Word. We have two opposing paradigms: Trust God and do good (Psa 37:3), or allow some amount of evil for the "greater good." "For the greater good" implies some children are expendable. This is not Biblical, or sane. If any parent knew that their own child would be one of the "expendable" children killed by this wicked worldview, they would never go along with this thinking. We must stand against the supposed "greater good" dogma to protect those who cannot protect themselves.

Sacrificing individuals "for the good of the society" is a logical fallacy in itself. Placing the importance of the individual before the society leaves the whole society stronger because the individuals who make up the society are stronger. Will a wall be more sound if each individual brick is as strong as possible, or if many of the bricks are compromised? Putting the good of the society first does not make society better, and this idea has been used as an excuse to allow terrible things in the name of "the greater good."

"On numerous occasions, from experiments by Nazi doctors to the U.S. Tuskegee Syphilis Study to the recent Pfizer trovafloxacin experiments in Nigeria, the modern world has observed the dangerous tendency to exact individual sacrifice in medical experiments for the alleged benefit of humankind."[7]

It is widely believed that infectious diseases disappeared because of vaccines, but we have seen that this belief is not based on historical facts. The truth is almost completely unheard of by most of the population. The argument "the end justifies the means" is also not valid if the means are more harmful than their (supposed) end. The United States government considers vaccines to be "unavoidably unsafe" products, defined as follows: "There are some products which, in the present state of human knowledge, are quite incapable of being made safe for their intended and ordinary use."[8] (They are not considered to be "unreasonably dangerous," either.) Why do we always hear vaccines are "safe and effective" when they are legally considered unsafe products?

Also, what about vaccine studies "proving" their safety? Well, we have already seen that a couple of standard claims about vaccines are untrue; it would be unwise to trust anything you have been told about them without further research. Here are some of the most striking (and perplexing) facts to help you. First and foremost:

"The medical community plays only a marginal role in determining the effectiveness and safety of vaccines. Instead, the vaccine industry, the very party that has a financial stake in the outcome, primarily drives safety verification, creating a huge conflict of interest."[9]

Another important factor: vaccine safety studies are usually done with placebos other than an inert substance. They are performed with an aluminum adjuvant as a "placebo," or sometimes a different vaccine is the "placebo."

"Vaccine adjuvants and vaccines may induce autoimmune and inflammatory manifestations in susceptible individuals. To date most human vaccine trials utilize aluminum (Al) adjuvants as placebos despite much evidence showing that Al in vaccine-relevant exposures can be toxic to humans and animals."[10]

"...the known toxicity of aluminium, is almost certainly a contributor to the success of aluminium-based salts as adjuvants and all that remains equivocal concerns whether this toxicity is confined to an area close to the injection site or if it is additionally manifested elsewhere in the body. None of this is controversial and all of this should be known and accounted for in any clinical trials of vaccines which incorporate aluminium-based adjuvants. So, why do so many trials use aluminium salts or unrelated aluminium adjuvanted vaccines as placebos for vaccinations which include an aluminium-based adjuvant? Not all vaccine trials use this ploy and instead use either an appropriate sham injection such as saline or the antigen in the absence of the adjuvant. It is necessary to make a very strong scientific case for using a placebo which is itself known to result in side effects and I have not found any scientific vindication for such in the recent human vaccination literature."[11]

If our children's health is the main objective of these "safety" studies, why is a false placebo used? Here are six studies that utilize aluminum adjuvant "placebos,"[12-17] which are toxic to the nervous system.[18] Many claim that because the amount of aluminum in vaccines is so small, and because we regularly ingest larger amounts of aluminum, it is therefore safe to inject a much smaller amount. As a counter to that statement, I would bring up venomous snakes. Snake venom is *venomous,* not *poisonous*; technically, it can be ingested. Our stomachs can handle the protein, break it down, and render it harmless (obviously, don't try this at home). If that same venom hits our veins, a completely different scenario will unfold. Ingestion and injection are not the same. Just because I can ingest something, does not mean I should inject it into my body, no matter how small the amount is, especially when it is a known neurotoxin.

Pediatricians shrug off the aluminum content they are injecting into our children because "it's such a small amount." Yet, a study from 2017 revealed something surprising: "An astonishing discovery about aluminum was made recently—less aluminum is worse. Less aluminum is *more* dangerous."[19]

"Aluminium (Al) oxyhydroxide (Alhydrogel®), the main adjuvant licensed for human and animal vaccines, consists of primary nanoparticles that spontaneously agglomerate. Concerns about its safety emerged following recognition of its unexpectedly long-lasting biopersistence within immune cells in some individuals, and reports of chronic fatigue syndrome, cognitive dysfunction, myalgia, dysautonomia and autoimmune/inflammatory features temporally linked to multiple Al-containing vaccine administrations. Mouse experiments have documented its capture and slow transportation by monocyte-lineage cells from the injected muscle to lymphoid organs and eventually the brain. The present study aimed at evaluating mouse brain function and Al concentration 180 days after injection of various doses of Alhydrogel®... An unusual neuro-toxicological pattern limited to a low dose of Alhydrogel® was observed. Neurobehavioural changes, including decreased activity levels and altered anxiety-like behaviour, were observed compared to controls in animals exposed to 200μg Al/kg but not at 400 and 800μg Al/kg. Consistently, microglial number appeared increased in the ventral forebrain of the 200μg Al/kg group. ... We conclude that Alhydrogel® injected at low dose in mouse muscle may selectively induce long-term Al cerebral accumulation and neurotoxic effects. ... In any event, the view that Alhydrogel® neurotoxicity obeys 'the dose makes the poison' rule of classical chemical toxicity appears overly simplistic."[20]

It appears the larger amounts of aluminum cause the body to respond more aggressively and contain the aluminum more effectively. It is unknown whether the long-term effects are better, worse, or the same. "This stunning realization should change the entire way we calculate safe levels of exposure to metal, especially in regard to aluminum."[21]

Notice that aluminum stays within the immune cells of some individuals. One author describes the implications of this phenomenon:

"If you have metals in your body, and something triggers your immune system to signal for help, it will inadvertently draw metals to that location. Metals and antibiotics in your body can cause what would have been a simple, localized bacterial infection to become systemic and prolonged. ... It's fairly obvious that when areas of your body are signaling for help but inadvertently get a toxin like aluminum, bad things can happen."[22]

Do you believe God intended for us to put this neurotoxic metal into our bodies via injection, bypassing the barriers He placed to prevent such a thing? Many more things can be said about aluminum; for now, suffice it to say, it should not be a "placebo" in vaccine safety studies.

Here are two studies that use a vaccine *as a placebo*.[23,24] One of these studies is testing to see if a certain vaccine increases the risk of *miscarriage*; the other is testing for the safety of using another vaccine on *adolescent girls*. What could possibly be the justification for this? If we as a society want to be absolutely sure something is safe for our children, why would we conduct safety studies in this manner? Why would *safety* studies use a *placebo* containing one of the most harmful components of the medication being studied? Would this not invariably skew the results? How can a clinical trial for an "unavoidably unsafe" product have another unsafe vaccine as the placebo? How is any of this even allowed? Because there are no laws when it comes to placebos.

"No regulations govern placebo composition. The composition of placebos can influence trial outcomes and merits reporting. ... Most studies did not disclose the composition of the study placebo. Disclosure was less common for pills than for injections and other treatments (8.2% vs. 26.7%). ... Placebos were seldom described in randomized, controlled trials of pills or capsules. Because the nature of the placebo can influence trial outcomes, placebo formulation should be disclosed in reports of placebo-controlled trials."[25]

In modern medicine, double-blind placebo-controlled studies are considered to be the very best, most reliable way to perform a study; yet vaccines are not subject to these studies.

"The best and most reliable form of research is the double-blind, placebo-controlled study. A treatment cannot really be said to be proven effective unless it has been examined in properly designed and sufficiently large studies of this type."[26]

"Vaccines are not subject to double blind clinical trials despite the evidence of vaccine-drug interactions and perhaps also of vaccine-vaccine interactions."[27]

Some claim these studies cannot be done because it is unethical to withhold vaccines from children. What is ethical about giving a false placebo? Supposedly, even letting someone go without a vaccine for a few days for a clinical trial is "unethical." But trial subjects are given a *different* vaccine, so they are still "unprotected" from this "vaccine preventable disease" that it was declared unethical to let them not be protected from in the first place, and the results of the study are still altered because a true placebo was not utilized. This makes absolutely no sense—it is as one author described, "asinine."[28] I'm not sure if there is a purported ethical reason to use an aluminum adjuvant as a placebo. It has been claimed that these false placebos are used to disguise vaccine reactions, and unfortunately, this appears to be true.

If the "most reliable form of research" is not used to test vaccines, and the trials performed usually do not use an actual placebo, how reliable are these trials? **Are these safety studies good enough for your children?** Here are some things to keep in mind regarding studies and clinical trials.

"...you can always find a study that contradicts everything else-or seems to. That doesn't mean that 'we don't know anything' or 'nothing is certain.' If you can find one hundred massive studies following thousands of people over many years that say one thing, and a single study covering only a few people for a few months saying the opposite, it's pretty easy to decide which evidence to believe."[29]

We must also recognize that bias and fraud are prevalent.

"Simulations show that for most study designs and settings, it is more likely for a research claim to be false than true. Moreover, for many current scientific fields, claimed research findings may often be simply accurate measures of the prevailing bias. … The greater the financial and other interests and prejudices in a scientific field, the less likely the research findings are to be true. Conflicts of interest and prejudice may increase bias. Conflicts of interest are very common in biomedical research, and typically they are inadequately and sparsely reported. Prejudice may not necessarily have financial roots. Scientists in a given field may be prejudiced purely because of their belief in a scientific theory or commitment to their own findings."[30]

"Although it might seem reasonable to assume that the incidence of misconduct or fraud in science in general and in clinical trials in particular is low, the true incidence is difficult to estimate for many reasons. … The under-reporting of fraud may be even more severe in medical research than in other scientific fields because of the negative consequences such reports have on pharmaceutical companies or other sponsors associated with fraud. …surveys of scientists and medical investigators routinely report rather high levels of misconduct or knowledge of misconduct. … Overall, the available evidence is rather unreliable but seems to suggest that the incidence of certain types of misconduct may be quite high but the incidence of data fabrication or falsification is low – but perhaps not as low as might be suggested by the frequency of confirmed and publicized cases. Given the large number of investigators involved, approximately 535,000 US scientists in 2012, even low percentages yield a large number of perpetrators."[31]

Vaccine safety studies are performed on healthy individuals, but the same people deemed unfit for these studies are readily recommended to receive those very vaccines once they are licensed.

"Physicians who conduct research on the pertussis vaccine will often eliminate what they consider to be high-risk children from their studies, even though these same categories of high-risk children in the general population are routinely given pertussis vaccine."[32]

"…vaccine testing is done on healthy subjects. But global dissemination after approval will include some of the sickest, most immune-suppressed people on the earth."[33]

Also, newborn children whose immune systems have not yet fully developed.

Bearing in mind the historical fact that vaccines did not cause infectious diseases to disappear, and what we've just seen about vaccine trials, let's look at some of the much less publicized research that associates harm with vaccine use. While not all vaccines are made using aborted fetal cell lines, the negative health effects of vaccination are not limited to just those vaccines; the vaccines that do not have human DNA will usually contain animal DNA. Additionally, many other aspects of vaccination can injure susceptible individuals. I wholeheartedly agree with the sentiments of Dr. Boyd E. Haley, who wrote:

"I truly wish I could write that I think vaccines are as safe and effective as the CDC claims they are, but I cannot. I have read hundreds of articles published in scientific journals, both supporting and questioning the safety and potential toxicity of vaccines, with many describing the basic biochemical and cellular toxic effects of vaccine components such as thimerosal and aluminum. Based on my extensive study of this issue over many years, as I write today, I have sincere and severe doubts about the safety and efficacy of vaccines (both those currently used and those that have been removed from the market) used to inoculate our children against infectious diseases. I am specifically concerned that current supporters of today's vaccine program quote as fact research results in articles proven to be based on manipulated data."[34]

Consider the following, starting from a child's earliest exposure to vaccines—before they are born.

Vaccines During Pregnancy

"We identified 132 reports of Tdap administered to pregnant women; 55 (42%) described no adverse event (AE). No maternal or infant deaths were reported. The most frequent pregnancy-specific AE was spontaneous abortion in 22 (16.7%) reports."[35]

It is recommended that pregnant women receive the Tdap (tetanus-diphtheria-acellular pertussis) vaccine to confer immunity to their newborns. I received this vaccine while pregnant, and now wish I hadn't, especially in light of the fact that "the strategy of vaccinating the household contacts of newborns (cocooning) failed to prevent pertussis among young infants."[36] **Why do we still recommend a failed strategy to expectant mothers with the documented common adverse event of spontaneous abortion?** The study concludes:

"During a time when Tdap was not routinely recommended in pregnancy, review of reports to VAERS in pregnant women after Tdap did not identify any concerning patterns in maternal, infant, or fetal outcomes."[37]

This is a technicality based on the fact that the vaccine *did not appear* to affect the children who were born; the children who died before birth were apparently not factored into this statement. This demonstrates the conclusions of studies cannot be trusted—many times they are misleading about the actual findings of a study.

Flu shots are also recommended to pregnant women. Here is a quote from one study showing that during the year two flu shots were recommended for pregnant women, the fetal loss reports on VAERS were much higher than usual.

"The ratio of the 140 fetal-loss reports during the incomplete 2009/2010 season to the 1.21 reports/year representing the mean of the 19 prior seasons, yields a 116-fold increase in fetal-loss reports in the VAERS database. ... Was the increase in fetal-loss outcomes during the two-vaccine 2009/2010 influenza season merely the result of reporting bias or was there a synergistic toxicity associated with the two-dose 2009/2010 influenza season? ... The observed reporting bias was too low to explain the magnitude increase in fetal-demise reporting rates in the VAERS database relative to the reported annual trends. Thus, a synergistic fetal toxicity likely resulted from the administration of both the pandemic (A-H1N1) and seasonal influenza vaccines during the 2009/2010 season."[38]

The overblown 2009 swine flu scare resulted in millions of unused H1N1 vaccines, which had to be incinerated because they had enough thimerosal to be considered hazardous waste.[39] The "116-fold increase" in fetal losses could certainly have been caused by such a toxic substance.

Here are some other reasons why pregnant women should avoid flu vaccines, *especially* those with thimerosal. Thimerosal has been removed from many vaccines (including all pediatric vaccines in 2001), and not all flu vaccines contain this compound, but:

"This is misleading. While most childhood vaccines no longer contain thimerosal as a preservative, thimerosal is still used in the manufacturing of vaccines and is still found in many vaccines in trace amounts. Mercury is a potent neurotoxin and any amount, even a trace amount, is worrisome, especially when injected."[40]

Pediarix™ is a vaccine administered to children that is claimed to be free of mercury but, according to the CDC's website, it contains "...a clinically insignificant trace"[41] of thimerosal. Even if a vaccine is completely free of thimerosal, many still contain aluminum, which

"...shares common mechanisms with mercury..."[42] (Thimerosal is about 50% mercury.) Mercury exposure during pregnancy has repeatedly shown to be associated with tics in children; and tics are associated with autistic disorders.

"We examined two primary exposure periods: the prenatal period and the period from birth to 7 months (1 to 214 days). We also evaluated exposures to mercury from hepatitis B vaccines and immune globulins in the first 28 days of life. ... We measured tics and stuttering dichotomously, and we estimated odds ratios for a 2-SD [standard-deviation units] increase in mercury exposure. ... The replication of the findings regarding tics suggests the potential need for further studies."[43]

"...hundreds of peer-reviewed studies...examine the effects of Thimerosal in animals and humans. Those studies link Thimerosal to a vast inventory of grim neurological disorders now epidemic in children, including ADD, ADHD, speech delay, tics, ASD, and autism. Other studies demonstrate Thimerosal's causation to asthma and a tenfold increase in Alzheimer's in adults exposed to Thimerosal laced flu shots. During three years of research, we were not able to find a single credible study in the open public health literature that shows Thimerosal to be safe. These impacts are no surprise. Mercury is a potent brain poison hundreds of times more toxic than lead. ... Thimerosal is so toxic that when a doctor carelessly shatters a multidose flu vial, state laws require evacuation of the building and clean-up by trained hazmat crews wearing protective boots, gloves, and respirators. Common sense should tell us that it's not a good idea to inject this poison into infants or expectant mothers. ... During my ten years observing this debate, I've been puzzled by the baseless insistence of media outfits and reporters that it is somehow safe to inject mercury into humans. ... Instead of arguing the science, they recite the credentials of the CDC's Vaccines Safety Division and repeat the CDC's hollow claim that 'The science shows Thimerosal is safe.'"[44]

Infant Mortality Rates (IMR)

Before reading the next quote, I want to remind the reader of the previously cited evidence in the last chapter concerning vaccination affecting the nervous system.

"When the cranial nerves are affected, the result is weakness of the eyes, ears, voice, and respiratory system, the latter condition being responsible for the cases of 'sudden infant death' following upon vaccination."[45]

It is argued that when hepatitis B vaccination for newborns became common practice, SIDS deaths decreased—this is another true but misleading statement from Paul Offit. After two measly paragraphs in a publication excerpted from one of his books, Offit concludes: "Therefore, the hepatitis B vaccine—like the pertussis vaccine—doesn't cause SIDS."[46] This assumption looks at no other information. The fact of SIDS deaths decreasing when hepatitis B vaccination increased is, again, technically true. But, the "Back to Sleep" campaign was also instituted during this time (which, to be fair, Offit does mention), and certain causes of SIDS *increased*.

"In 1992, to address the unacceptable SIDS rate, the American Academy of Pediatrics initiated a 'Back to Sleep' campaign, convincing parents to place their infants supine, rather than prone, during sleep. From 1992 to 2001, the postneonatal SIDS rate dropped by an average annual rate of 8.6%. However, other causes of sudden unexpected infant death (SUID) increased. ... A closer inspection of the more recent period from 1999 to 2001 reveals that the US postneonatal SIDS rate continued to decline, but *there was no significant change in the total postneonatal mortality rate.* During this period, the number of deaths attributed to 'suffocation in bed' and 'unknown causes,' increased significantly."[47]

"There seems to be little investigation into the SIDS phenomenon, in spite of the fact that it is the leading cause of death in babies born healthy!"[48]

There is no investigation because those at the very top know what causes SIDS, and they don't want the public to know (see Isa 32:6-7). The claim that we "just don't know" what causes SIDS, but it can't be vaccines, is not based on a critical review and analysis of the available scientific information. Paul Offit's two paragraphs (which are from an excerpt, maybe the actual book contains more information) and conclusion are woefully inept on this heartbreaking tragedy that affects thousands of families every year. Babies are not born to die for "no reason." While there probably are some rare cases that are truly "unexplained" because of man's limited knowledge, there is a reason these babies are dying; and there is much evidence that it is largely due to the formidable assault on their delicate bodies in the form of vaccination.

In 1982, William C. Torch presented a paper at the Thirty-Fourth Annual Meeting of the American Academy of Neurology. His conclusion stated that DPT "...may be a generally unrecognized major cause of sudden infant and early childhood death, and that the risks of immunization may outweigh its potential benefits. A need for re-evaluation and possible modification of current vaccination procedures is indicated by this study."[49] Torch was "...vigorously criticized for his 'non-prospective' study, his use of 'anecdotal data,' and his conclusions."[50] There were studies published several years later which found no link between DPT and infant death. There have to be "studies" to restore trust in the immunization program; doubts cannot be allowed to exist.

Case studies also claim no link exists between SIDS and vaccines, but many cases did not require autopsies. Nearly half of the children in one study did not have an autopsy, yet the study "...was cited in 2005 as a 'proof' of absence of association between SIDS and vaccines."[51] Another study concludes there was no association between SIDS and DPT; however, it doesn't give any figures that have not been "controlled"—just the adjusted conclusions that claim there is no link. This study references many other studies, and the only issue examined is DPT and SIDS—not vaccination in general. It mentions the different methods the other cited studies utilized and states that not including completely unvaccinated children is a benefit of a certain study "approach" because "biases" are avoided.

"Because this approach is limited to exposed cases only, that is, those children who receive DPT vaccination and die of SIDS, the power to detect an elevated risk <u>is lower</u> than in the first approach. On the other hand, <u>potential biases arising from inclusion of unvaccinated children are avoided</u>."[52]

Notice, this method isn't as likely to find a problem if there is one, which is supposedly the point of these studies, but hey, certainly no "biases" from having a legitimate control group with no vaccinations. How about we just include all of the available information so we have reliable studies? There could be an unlimited number of studies concluding there is no link between SIDS and vaccination, but they could not convince me to ignore the evidence I have seen to support the opposite. This report published by *Pediatrics* was already mentioned, but it is highly significant. So, again: SIDS deaths *disappeared* in Japan when they postponed their vaccination age. Japan's schedule has changed since this report, but even today, according to the CIA's website, Japan has the second lowest infant mortality rate in the world. (This study specifically examines DPT and DTaP; whether Japan postponed other immunizations as well is not stated.)

"The category 'sudden death' is also instructive in that the entity disappeared following both whole-cell and acellular vaccines, when immunization was delayed until a child was 24 months of age."[53]

Why did the American Academy of Pediatrics not stop DPT vaccination, or at least inform parents of the risks, or at the very least investigate further? Is Offit, who has stated that the pertussis vaccine does not cause SIDS, unaware of this study? Probably not. There are two options: he is vastly ignorant on a subject that the public is paying him *to educate them* about, or he is outright lying about the pertussis vaccine, its role in infant death, and is knowingly telling parents **a lie that could result in the deaths of their children**. Either one means he is not a trustworthy source of information about vaccines.

This information has been known for decades, and as we saw in the last section pertussis mortality actually *goes down* when pertussis vaccination rates are lowered. How many children have died every year because of "SIDS" after receiving a vaccine that did not even protect them from whooping cough? It has also been known for decades that the risks of whooping cough vaccination outweigh its "benefits." The medical establishment has repeatedly failed to warn parents of the dangers of vaccination.

"The DPT shot had become a routine medical procedure in America; and, instead of being informed about vaccine reactions, parents were legally required to have their children vaccinated in order to enroll them in school. When a child reacted violently to a DPT shot, parents were rarely told it was the pertussis vaccine that had caused their child to die or become brain damaged. Not until reporter Lea Thompson...broke the story in April 1982 with the documentary 'DPT: Vaccine Roulette' were American parents made aware that their children could die or be permanently injured from a pertussis vaccination. It had been thirty-four years since Byers and Moll* had informed the American medical establishment of that fact."[54]

(*The Byers and Moll study from 1948 stated that encephalopathy following pertussis vaccination was more devastating than the lesions caused by the smallpox vaccine; it was cited in the previous chapter.)

SIDS was a concern brought up to my child's (former) pediatrician, because the time that a child is most likely to die in their sleep for "no reason" coincides with the CDC's immunization schedule. This was regarded as coincidence, as being unrelated, and many studies declare the same conclusion. **It is baffling *pediatricians* are ignorant of an entire country's change of immunization policy which drastically lowered their SIDS rates.** Thousands of studies stating there is no link between vaccination and SIDS would not make me ignore **what actually happened** and was observed on a **nationwide scale** in Japan: postponing the age of immunization caused these "sudden" deaths to disappear. (They can't be called "unexplained deaths" when the explanation is obviously right there.) There really is not much to doubt about vaccination causing SIDS.

"It would be just too much of a coincidence that tens of thousands of babies die after DPT injections, yet none of these deaths would be caused by such a highly noxious substance as DPT vaccine. This is especially instructive since the vast majority of these babies often were healthy bouncing infants just hours before they were injected with DPT and died."[55]

"The diphtheria-pertussis-tetanus (DPT) vaccine control officer called the DPT shot then in use the 'dirtiest material put into humans.'"[56]

"The incidence rate ratio of the 1999-2001 rate to the 1989-1991 rate is 1.26 indicating a statistically significant increase in proportion of neonatal SIDS since implementation of the universal vaccination of newborns against hepatitis B and gradual increase in coverage by the birth dose."[57]

"It took almost three years of researching into babies' breathing before we looked at each other and said 'Vaccines are killing babies.'"[58]
Viera Scheibner, PhD, from "Vaccination the Hidden Truth" Documentary

"Knowledge of newborn immune systems is inadequate; the duration of immunity, if any, conferred by the vaccine is totally unresearched and there is a wealth of information on neurological side-effects following vaccination..."[59]

"The U.S. requires infants to receive 26 vaccine doses, the most in the world, yet 33 nations have better infant mortality rates. ... Developed nations that require the least number of infant vaccines tend to have the best infant mortality rates."[60]

"Of the 34 nations that have crossed the socio-economic threshold and are able to provide the basic necessities for infant survival—clean water, nutrition, sanitation, and health care—several require their infants to receive a relatively high number of vaccine doses and have relatively high infant mortality rates."[61]

Notice, the United States' IMR is the worst of all other developed countries, *and* gives the most vaccines.

"...IMR and vaccine doses tend to increase together. ... It appears that at a certain stage in nations' movement up the socio-economic scale—after the basic necessities for infant survival (proper nutrition, sanitation, clean water, and access to health care) have been met—a counter-intuitive relationship occurs between the number of vaccines given to infants and infant mortality rates: nations with higher (worse) infant mortality rates give their infants, on average, more vaccine doses. ...unvaccinated babies who died of SIDS did so most often in the fall or winter while vaccinated babies died most often at 2 and 4 months—the same ages when initial doses of DPT were given to infants. ...babies died at a rate nearly eight times greater than normal within 3 days after getting a DPT vaccination. ... A closer inspection of correlations between vaccine doses, biochemical or synergistic toxicity, and IMRs, is essential. All nations—rich and poor, advanced and developing—have an obligation to determine whether their immunization schedules are achieving their desired goals."[62]

In the United States, we no longer use the whole cell DPT but instead the acellular DTaP and "booster" Tdap. However, it is important to note this issue is not about any one ingredient or any one vaccine. It is about the fact that too much is given too soon, based on a schedule influenced by vaccine patent owners[63] (see excerpt with source information), in an industry wrought with questionable practices and fraud, and reasoning built upon lies. This vaccine schedule starts with pregnant women receiving vaccines, through the childhood schedule that amounts to about 70 doses.[64] The vaccine schedule our parents/grandparents had is much different than what children have today. Yes, some vaccines are less dangerous than they were, but we cannot expect a schedule of a handful of shots versus almost 70 to not have different effects. In the 1950s, the United States "...had the third best infant mortality rate in the world..."[65] ***What has been the biggest universal change in pediatric medicine coinciding with the drastic increase in infant mortality?***

"In the early 1950s, four vaccines were available: diphtheria, tetanus, pertussis and smallpox. Because three of these vaccines were combined into a single shot (DTP), children received five shots by the time they were 2 years old and not more than one shot at a single visit. By the mid-1980s, seven vaccines were available: diphtheria, tetanus, pertussis, measles, mumps, rubella and polio. Because six of these vaccines were combined into two shots (DTP and MMR), and one, the polio vaccine, was given by mouth, children received five shots by the time they were 2 years old and not more than one shot at a single visit. Since the mid-1980s, many vaccines have been added to the schedule. The result is that the vaccine schedule has become more complicated than it once was, and children are receiving far more shots than before. Now, children could receive as many as 27 shots by 2 years of age and up to six shots in a single visit."[66]

Here are studies on unexpected deaths after hexavalent vaccination.[67,68]

"...a 16-fold risk increase after the 4th dose could be detected with a power of at least 90 per cent. A general 2-fold risk increase after vaccination could be detected with a power of 80 per cent."[69]

We have all seen and heard drug commercials that list side effects from that *one* medication. Imagine the list of side effects from just one vaccine—something injected rather than ingested. What about a *pregnant woman* receiving multiple vaccines? Imagine the commercial for the hepatitis B vaccine given to *a newborn child* before he or she leaves the hospital, or for the many shots received at just two, four, six, nine, and twelve months of age. Imagine the list of side effects for the entire vaccination schedule. Vaccines have never been tested for their cumulative effects on health, as we will see soon. With all you have learned about vaccines so far, is this really what you would want for your child?

"Other developed countries vaccinate much more cautiously than we do in the United States. Their citizens are not dying from infectious diseases. As opposed to the thirty-seven injections we have before age six, other developed countries give far fewer injections: Iceland (11), Sweden (11), Singapore (13), Japan (11, and pulled their MMR vaccine due to high injury rate), Norway (13), Hong Kong (13), Belgium (18), Austria (19), Israel (11), Denmark (12), Netherlands (20) [in 2015]."[70]

Regarding SIDS, there is one other subject on which parents should be informed. In his book, *Every Second Child,* Dr. Archie Kalokerinos tells of the many infants saved from what would have been considered unexplained infant deaths by the simple administration of vitamin C. His research and observations led him to notice that children with mild illnesses would collapse and die, or be left in their cribs and later found dead. Vitamin C was even administered shortly after some of his patients collapsed, and they quickly recovered. His observations were met with disbelief and scorn by almost all of his colleagues. His patients were Australian Aboriginal children who were immunologically predisposed to infections, and had a vitamin-deficient diet which caused many of them to die of unrecognized infantile scurvy. We will read an excerpt from his book, but first, I want to point out something from the Scriptures.

Genesis 1:29 **And God said, Behold, I have given you every herb bearing seed, which is upon the face of all the earth, and every tree, in the which is the fruit of a tree yielding seed; to you it shall be for meat.**

Biblically, and botanically, any produce that contains its seeds is a fruit; things like tomatoes, peppers, and avocados are technically fruits. The foods with the highest vitamin C content are also fruits. If we still ate according to the instructions that God gave Adam and Eve in the garden of Eden, we would ingest large amounts of this vitamin. I presume I will be forgiven for mentioning vitamin C so much: to understand why, one needs only to read a few accounts of the seemingly miraculous recoveries of ailing individuals thanks to the administration of this vitamin found in abundance in the original diet that God designed for man. Notice this account:

"Hardly any system in the body can function without adequate amounts of vitamin C. During periods of deficiency, serious biochemical and immunological problems may become manifest; then the reason for an infection that usually precipitates a deficiency through increased utilization must be considered. ... The reason why sudden deaths of this nature do not occur in older children or adults is simple biochemistry. The variations of an infant's requirement of vitamin C are tremendous. Adults do not exhibit this. Adult scurvy is therefore a slow process. Serious signs of deficiency occur before death takes place. A few months after I began my research into sudden infant deaths I was able to demonstrate to a colleague the reversal of the shock stage with vitamin C. Once a little boy walked into my consulting room with his mother. He came from a distant town because, as his mother explained, 'The local doctor does not treat the children as you do.' An examination revealed an acute upper respiratory tract infection and swollen, red gums typical of infantile scurvy. There was no doubt that the boy was seriously at risk—even though he seemed perfectly happy. I admitted him to the hospital, but because my colleague was anxious to measure the amount of vitamin C in the urine before treatment was commenced, I gave instructions that the injection was to be withheld until a specimen was collected. The boy was slow in obliging. Two hours went by and a specimen had not been collected, Then, without further warning, there was collapse and shock. Of course I did not wait any longer for a specimen. The vitamin was injected quickly. The effect was excellent. Within half an hour the boy was well."[71]

Pregnant and nursing women need to make sure they consume enough vitamin C. While modern medicine says SIDS cannot be explained, we know a link between smoking and SIDS quite possibly exists due to the exposure to toxins from cigarette smoke; exposure to toxins after vaccination utilizes vitamin C as well, possibly leaving a child deficient. Vaccination is known to cause an "...increased utilization of vitamin C..."[72] Dr. Kalokerinos fought to make this information known in an attempt to save children's lives. When the infant death rates doubled in some areas, reaching rates as high as **500 per 1000**, he realized the rise coincided with the implementation of more aggressive vaccination campaigns. He knew that increasing immunizations would cause further depletion of vitamin C levels among this already scorbutic population, and this is what had caused the death rate to increase. His theory was verified in an animal experiment:

"One research worker in the laboratory had been immunizing animals against diseases like tetanus and diphtheria. His experiences showed that after being immunized, some of the animals died suddenly within twenty-four hours. These deaths had been attributed to anaphylaxis. Authorities the world over had decided that this was so (it is a severe allergic reaction). I suggested that vitamin C deficiency was the cause. The animals involved did not make their own. Like primates they required it in their diets. To discover the truth required only a simple experiment. In Ames Laboratories this had been done. The result was definite, unquestionable and final. Half of a group of animals were supplemented with vitamin C before being immunized. None died. The unsupplemented half continued to die at rates equal to those found in previous experiments. The importance of this discovery can hardly be stressed. In Australia and all over the world. Infants were being immunized. Those whose vitamin C status was low were at risk. Here, at last, was experimental evidence that supported my claims that stepping up immunization campaigns among Aboriginal infants increased the death rate. Many died from acute vitamin C deficiency precipitated by the increased utilization of vitamin C following the immunization. There was, of course, also the problem of immunological insults resulting from the immunization in susceptible infants that make them particularly prone to serious infections. The Miles experiment did not involve this aspect."[73]

While a vitamin C deficiency cannot unequivocally explain every tragedy of an unexplained infant death, if the medical community refuses to examine or **investigate to see if a completely safe vitamin could save thousands of lives every year**, continued ignorance is certain.

Vaccination can cause encephalitis—as there are studies which state pertussis toxoid is used to induce encephalitis. It has also been demonstrated that vitamin C has cured patients affected by encephalitis; many babies exhibit symptoms of encephalitis before dying unexpectedly.

"The child who died 20 days after the second absorbed DPT had been unwell since vaccination with attacks of high pitched crying 3 or 4 times a day for 10 days, followed by a period of excessive sleepiness. The high-pitched crying recurred 4 days...before the baby's death on the 20th day. ... Many cases of screaming, limpness, pallor, glazed eyes etc. were reported by these authors within 2 hours after both DPT and DT vaccine administration. ... There is not a shadow of doubts that vaccines are killing and maiming babies."[74]

Why are these babies screaming? Because they are in terrible pain due to encephalitis? Chances are, if one called their child's pediatrician to report high-pitched cries after vaccination, the child's crying would be considered a normal occurrence. Something is happening to these children that is causing them to scream in pain. There have even been reports of "...bulging fontanelle associated with increased intracranial pressure..."[75] after vaccination. I agree with Viera Scheibner, who wrote: "The total oblivion of doctors to encephalitogenic effects of vaccines administered to young babies is quite incredible, to say the least."[76] In their defense, doctors have always been educated to remain ignorant; and with studies that supposedly dismiss vaccines of having any correlations with negative effects, like "'Pertussis Vaccine Encephalopathy': It Is Time to Recognize It as the Myth That It Is" they don't just stop at ignorant but become fully indoctrinated to be believe there is nothing to question when it comes to vaccines. *Doubts cannot be allowed to exist.* There are several studies which declared pertussis toxoid is used to cause encephalitis in animal experiments; it's no myth that this same toxoid could cause encephalitis in infants.

Today, it is acknowledged vaccines can cause encephalitis, but many times when children are clearly having an adverse reaction to a vaccine and displaying symptoms of encephalitis, parents are told this is normal. If an adult exhibited these symptoms, "...they are readily and without hesitation described as signs of encephalitis or encephalopathy."[77] If parents were given true informed consent, they would rarely choose to let their children receive a vaccine that carried such risks, especially when one realizes the low efficacy of this vaccine.

When reading accounts of healthy babies dying shortly after their DPT injections, it's amazing how medical professionals are quick to say the shot had nothing to do with the death of the baby, even though the baby was declared to be in perfect health right before administration of the vaccine. Parents have been awarded compensation following the deaths of their infants after DPT vaccination, and many of the death certificates of these children have SIDS as their cause of death.

"In January 1990, the United States Claims Court revealed that 92 cases of vaccine injury and death had been decided under the National Childhood Vaccine Injury Act of 1986. More than half of the awards were made to parents whose children had died following DPT vaccine reactions. Most of the children's death certificates listed the cause of death as 'Sudden Infant Death Syndrome.'"[78]

It has been said the reason infants are vaccinated is because they cannot communicate that a vaccine has harmed them, except by displaying some of the above-mentioned symptoms, which are considered "normal."

"The reasons for this most unfortunate interpretation are that vaccines are considered innocuous and totally safe and that babies cannot speak. A baby screaming in pain is usually ignored or handled almost with contempt."[79]

One five-year-old boy had a vaccine reaction, and he was capable of describing his pain:

"Because my son was five when he reacted, he was able to verbalize what he was feeling. He could just as easily have reacted when he was two or four months old and would have been unable to tell me how he was feeling. I didn't even get him home from the doctor's office before he started writhing on the floor of the car, screaming that he had a headache. By the time we got home, his fever had reached 104 degrees.

He vomited, ran around the house, and hid in corners as if he was frightened to death."[80]

This is not "normal." Scheibner also made the following comment (of which I am in complete agreement):

"Tests on animals are much less valuable because animals can't tell us how they feel. One of the best exercises of this kind would be, for the doctors administering vaccines to inject themselves with the same vaccines as routinely administered to babies (and in doses in a proper proportion to body weight). They would soon know how babies feel and what effect the vaccines have on them."[81]

"...hell-bent on vaccination, irrespective of side-effects, the vaccinators are... The fanaticism fed by ignorance, irrational fear of illness and greed are the moving forces behind the ritual of vaccination. Vaccine injections represent an enormous and unjustified insult to a young child. ...despite all the demonstrated ineffectiveness and dangers of various vaccines, vaccination must go on. The pain and deaths of babies seem quite irrelevant. As if the death of a small baby was a lesser death."[82]

My child didn't receive one vaccine her first year of life; when I put her to bed, after taking precautions to make sure she couldn't suffocate, I didn't spend one second worrying about her dying "unexpectedly." My sleep was sweet, and I was not afraid of "*sudden fear*" (Pro 3:24-25). I was convinced that vaccines cause almost all SIDS deaths; I have not learned a single thing that makes me doubt this conviction.

"No amount of denial will cover up thousands of pages of papers reporting adverse reactions, including encephalitis and death, reported in medical journals over the decades of mass vaccination. It would be straining credulity too much to assume that thousands and thousands of cases of brain damage and tens of thousands of deaths following DPT vaccination were all coincidental."[83]

"Every biological event occurs for a reason, even if it is not commonly known or appreciated."[84]

...Or properly investigated. Babies do not die for no reason; this does not "just happen." Encephalitis caused by vaccination can damage a child's cranial nerves and affect their breathing, resulting in death.

"The lungs and respiration are under direct control of the nervous system, and breathing difficulties plus a bluish tint of the skin or pallor have often been reported by parents of infants who collapsed after a DPT shot."[85]

"Breathing difficulties such as SIDS and asthma are due to impairment of the cranial nerves governing respiration. ...cranial nerve palsies are the physiologic substrate of postencephalitic syndrome."[86]

"Harvey Jackson, Jr., is a classic case. According to his father's diary, at age three months his breathing stopped altogether—thirty-six hours after his second DPT shot. The father describes this occurrence: ... 'A difficult and frightening day, Harvey had been wheezing loudly since the morning, and by one to two P.M. he was so severely tried in his breathing that I became terribly alarmed. He had to be held upright so he could breathe. Once, possibly twice, as I held him, he stopped breathing, only to start against after being thrown over my shoulder and hit on the back.' ... Harvey, Sr., believes that Harvey, Jr., narrowly escaped 'sudden infant death.' The boy grew up to be moderately autistic. Thus autism and SIDS appear to be generated in the same way—by an encephalitis most commonly caused by vaccination. Breathing difficulties of various kinds ensue, and during one such 'breath-holding attack' the infant simply expires. The child who would have grown up autistic dies instead of 'sudden infant death syndrome.'"[87]

Autism doesn't "just happen," either.

Autism

"This nationally demonstrated crisis, using the government's own data on infant mortality, defies any claim of vaccine safety by the CDC. To the contrary, this exceptionally high death rate supports the concept that the CDC vaccination program is causing severe damage to our infants who live. ... While some studies do not find evidence of an association between vaccines, heavy metal components such as thimerosal, and autism, many do. The peer-reviewed meta-analysis released by DeSoto and Hitlan...found that 74 percent of the relevant studies support an association between autism and heavy metals such as thimerosal. There are now [in 2012] approximately sixty studies that support the autism-vaccine causation theory. However parents, attempting to make fully informed decisions about vaccinating their children who Google 'Do Vaccines cause autism spectrum disorders?' will see the page containing the CDC's inaccurate claim as the first search result that is returned to them. The trust those parents have that the government is giving them accurate information is misplaced. ... I, and other involved researchers, would gladly debate these issues in public with CDC and other experts—if they had the courage to show up. They don't."[88]

I want to give some quick comments on the studies concluding there is no link between MMR or mercury and autism. First, there needs to be a large-scale study looking at completely unvaccinated versus vaccinated children to get a true measure of this issue. Studies examining just one ingredient or only one vaccine are missing the big picture—whether or not vaccines can trigger the development of autism in susceptible individuals. As we've just read, most of the science actually shows there is a link between vaccines and autism. One of the main researchers, Poul Thorzen, an author on some of the autism studies which conclude there is no link, is now wanted for many counts of fraud. If you search his name, the government website where you can see his fugitive profile is the first result. We have been told many, many times that "vaccines do not cause autism." Let's see if the science is truly as "settled" as it is claimed to be.

Dr. William Thompson was one of the authors of the main studies considered to be the cornerstone of the CDC's science to supposedly prove vaccines do not cause autism. He became a CDC whistleblower and stated that an MMR autism signal was concealed in their studies. These excerpts are from a book titled *Vaccine Whistleblower: Exposing Autism Research Fraud at the CDC*, followed by an excerpt from a published study:

"[Dr. William] Thompson is an author of two of the three epidemiological studies on American populations touted by the CDC to 'prove' the safety of Thimerosal. He is also coauthor of the CDC's seminal 2004 study known as DeStefano 2004, which dismissed the link between the MMR vaccine and autism. That study has been cited in ninety-one subsequent published studies and is one of the principal cornerstones for claims by the CDC and the vaccine industry that vaccines do not cause autism. Thompson now confesses that he and his fellow CDC researchers found a strong autism signal in children who received the MMR vaccine before their third birthday. Then, under orders from their bosses and in violation of the study protocols, the scientists eliminated this data from the final published study, in order to fool the public about vaccine safety. ... He now admits that, under pressure from his superiors, his team fraudulently withheld data demonstrating a 340 percent higher risk of autism in African American boys who received that vaccine on time compared to the boys who delayed the vaccine. ... On the basis of all the population data and the CDC's most recent autism incidence estimates, at least 100,000 African American male children could have been spared debilitating neurological injury if the CDC scientists had told the truth when Thompson and his team first discovered the increased risk in 2001."[89]

"Autism prevalence is up more than 14,000 percent over the thirty-year period from 1985 to 2015. This increase eviscerates the theory that genes cause autism. There is no such thing as a genetic epidemic. The great autism gene hunt has been a failure. The dramatic increase of autism over the past twenty years must be attributable to something in the environment. Therefore, it is necessary to look for causative or contributing environmental factors. Based on tens of thousands of parental reports of children regressing into an autism diagnosis after vaccination, vaccines were a prime suspect. The CDC and the Institute of Medicine [IOM] went through a charade of investigating the vaccine schedule to see if vaccination had anything to do with the increase in autism. Rather than looking at all vaccines and the schedule as a whole, they purported to investigate ONE vaccine (MMR) and ONE vaccine component (thimerosal). Based on the superficial review, they declared in 2004 that the entire vaccine schedule was safe, and then thwarted additional research into vaccine safety going forward, as stated in the 2004 IOM Vaccine Safety Review. Dr. William Thompson, a coauthor on MMR and thimerosal studies, now says the CDC has wasted ten years by failing to do credible research, even with respect to MMR and thimerosal—the only two aspects of the vaccine schedule that were looked at in the superficial review!"[90]

"In the past ten years, more children have regressed into autism. Their suffering could have been avoided. Lives and families have been ruined. A week does not pass without a report of a child with autism drowning, getting hit by a car, or being abused."[91]

"A 2007 US Senate investigation by Senator Tom Coburn, 'CDC Off Center,' found widespread corruption and mismanagement at the CDC's vaccine programs."[92]

"Epidemiological or population studies are well-suited to arriving at prearranged results. ... That's the gimmick the CDC has perfected, in order to preserve the illusion of Thimerosal safety. In the 2007 study, CDC scientists removed the low IQ individuals with autism or other neurological diagnosis from the pool before even beginning their study on Thimerosal exposure."[93]

"In 2004, the US Center for Disease Control (CDC) published a paper showing that there is no link between the age at which a child is vaccinated with MMR and the vaccinated children's risk of a subsequent diagnosis of autism. One of the authors, William Thompson, has now revealed that statistically significant information was deliberately omitted from the paper. Thompson first told Dr S Hooker, a researcher on autism, about the manipulation of the data. Hooker analysed the raw data from the CDC study afresh. He confirmed that the risk of autism among African American children vaccinated before the age of 2 years was 340% that of those vaccinated later."[94]

In the documentary, *Vaxxed: From Cover-Up to Catastrophe*, a pediatrician named Jim Sears, MD, is shown the information that was fraudulently concealed by the CDC and said this with incredulity:

"Everything I've been telling my patients for the last ten years has been based on a lie and a cover-up." Why is it that the media, which supposedly stands in solidarity with the Black Community, does not cover this story? How many black men will go through life handicapped because of this government agency?

In one autism/MMR study, many children did not receive their second MMR dose until twelve years of age—much later than the current US schedule. While the study concluded there is no link between MMR and autism, it states: "The highest risk for autism was conferred by being a boy, being born in a late birth cohort (2008-2010)..."[95] Earlier in the study it is also stated: "A first dose of MMR vaccine is offered at 15 months, with a second dose at 12 years of age or, since 2008, at 4 years of age."[96] According to this study, one of the highest risks of being autistic was being in the later birth cohort, which received the MMR vaccine eight years earlier than the other cohort. So, it would appear that getting the MMR earlier actually was associated with an autism diagnosis; the conclusion doesn't seem to match the evidence. This study also examined children much too young for an autism diagnosis and had other issues that seemed odd, even to my uneducated eyes. There is an online article titled "A Scientists Rebuttal to the Danish Cohort Study," where eight issues with this study are highlighted and briefly explained. It concludes: "With these issues, this paper cannot be relied upon as evidence that the MMR vaccine does not cause autism."[97]

I will admit, I haven't read many of these studies that show no link between autism and MMR. But, the studies I read do not leave me with the impression that everything possible is being done to make sure there is absolutely no association between vaccines and autism; they appear more as though they are designed not to find any such link. Remember, the US government stated over thirty-five years ago that "...doubts...about the safety of the vaccine cannot be allowed to exist..."[98] Maintaining the public's trust in immunization is the goal, otherwise, their house of cards comes crashing down.

Now, let's look at some studies that show the opposite of what the CDC claims and what the mainstream media repeats regarding vaccines and autism.

"A growing body of scientific literature points towards the contribution of vaccines to rising childhood immune and neurological disorders."[99]

"We have examined the neurotoxicity of aluminum in humans and animals under various conditions, following different routes of administration, and provide an overview of the various associated disease states. The literature demonstrates clearly negative impacts of aluminum on the nervous system across the age span. ... In young children, a highly significant correlation exists between the number of pediatric aluminum-adjuvanted vaccines administered and the rate of autism spectrum disorders."[100]

"The reason for the rapid rise of autism in the United States that began in the 1990s is a mystery. Although individuals probably have a genetic predisposition to develop autism, researchers suspect that one or more environmental triggers are also needed. One of those triggers might be the battery of vaccinations that young children receive. Using regression analysis and controlling for family income and ethnicity, the relationship between the proportion of children who received the recommended vaccines by age 2 years and the prevalence of autism (AUT) or speech or language impairment (SLI) in each U.S. state from 2001 and 2007 was determined. A positive and statistically significant relationship was found: The higher the proportion of children receiving recommended vaccinations, the higher was the prevalence of AUT or SLI. A 1% increase in vaccination was associated with an additional 680 children having AUT or SLI. Neither parental behavior nor access to care affected the results, since vaccination proportions were not significantly related (statistically) to any other disability or to the number of pediatricians in a U.S. state. The results suggest that although mercury has been removed from many vaccines, other culprits may link vaccines to autism. Further study into the relationship between vaccines and autism is warranted."[101]

"Aluminum also shares common mechanisms with mercury e.g. it interferes with cellular and metabolic processes in the nervous system. Children given the recommended vaccinations are injected with nearly 5 mg of aluminum by the time they are just 1.5 years old, almost 6 times the safe level. Furthermore the nature of the Aluminium affects the prevailing blood levels and is also increasingly implicated, through their use as vaccine adjuvants, in autism."[102]

"Aluminum (Al), the most commonly used vaccine adjuvant, is a demonstrated neurotoxin and a strong immune stimulator. Hence, adjuvant Al has the potential to induce neuroimmune disorders. When assessing adjuvant toxicity in children, two key points ought to be considered: (i) children should not be viewed as "small adults" as their unique physiology makes them much more vulnerable to toxic insults; and (ii) if exposure to Al from only few vaccines can lead to cognitive impairment and autoimmunity in adults, is it unreasonable to question whether the current pediatric schedules, often containing 18 Al adjuvanted vaccines, are safe for children? ... Our results show that: (i) children from countries with the highest ASD prevalence appear to have the highest exposure to Al from vaccines; (ii) the increase in exposure to Al adjuvants significantly correlates with the increase in ASD prevalence in the United States observed over the last two decades; and (iii) a significant correlation exists between the amounts of Al administered to preschool children and the current prevalence of ASD in seven Western countries, particularly at 3-4 months of age."[103]

"The mass of scientific evidence compiled by researchers clearly indicates that the incidence of autism occurs following vaccination and is most closely associated with the schedule of vaccines culminating in the MMR vaccine. ... By contrast Autism appears not to occur in communities which do not use vaccines"[104]

Despite the dogma that vaccines do not cause autism, there are many cases in which compensation was given to families of vaccine-induced autistic children through the government's National Vaccine Injury Compensation Program. One study reviewed these cases and revealed the following:

"This study found eighty-three cases of acknowledged vaccine-induced brain damage that included autism. In twenty-one published cases of the Court of Federal Claims, the Court stated that the petitioners had autism or described autism unambiguously. In sixty-two additional cases, the authors identified settlement agreements where HHS [Department of Health and Human Services] compensated children with vaccine-induced brain damage as well as autism or an autism spectrum disorder. ... This finding of autism in compensated cases of vaccine injury is significant, as U.S. government spokespersons have been asserting the absence of a vaccine-autism link for more than a decade."[105]

Honestly, when it comes to vaccine side effects, the list appears to be unending. As Dr. Robert Mendelsohn aptly wrote on this matter: "The subject is so vast and complex that it deserves a book of its own. Consequently, I must be content here with summarizing my objections to the fanatic zeal with which pediatricians blindly shoot foreign proteins into the body of your child without knowing what eventual damage they may cause."[106] I will list additional sources at the end of this book for anyone who wishes to research further.

Next, I want to point out the following: vaccines damage DNA and alter the immune system. Many of the negative effects of vaccination can be traced back to these two main facts.

"Many scientists have now hypothesized some cancers and other childhood neurological disorders such as Autism Spectrum Disorder (ASD) may be caused by mutations in the child's DNA. This conclusion is based upon finding numerous mutations in the brains and/or blood of such children which cannot be found in their parents. Some mutations occur at or soon after conception and are called de novo. Others are acquired mutations that occur after birth due to environmental factors or exposures and are called somatic mutations. One such event could be exposure to residual human DNA from childhood vaccines."[107]

"Recombinant virus vaccines that express a limited number of epitopes are currently being developed to prevent disease by changing the relative balance between viral spread and the immune response. Some circumstances, however, were found in infections with a noncytopathic virus in which vaccination caused disease; sensitive parameters included the genetic background of the host, the time or dose of infection, and the constituents of the vaccine. Thus, immunopathologic damage by T cells may be an unwanted consequence of vaccination with the new types of peptide or recombinant vaccines that are being investigated for the human immunodeficiency viruses and other pathogens."[108]

"This article explores the issues and concludes that sensory dysfunction and systemic failure, manifested as autism, is the inevitable consequence arising from subtle DNA alteration and consequently from the overuse of vaccines. ...

Modified live viruses alter the structure and function of DNA. Each virus is a large molecule therefore its spatial arrangement must be influenced by its biochemistry which influences cross-helical structures and linkages within the DNA helix. Accordingly it is inevitable that the steady accumulation of such foreign proteins arising from an intensive vaccine programme will reach the stage where it significantly weakens DNA, gene, and chromosome structure and function. ...

The introduction of modified live viruses as vaccines enable the virus to attach its genetic material into the cell which replicates i.e. the host cell continues to function whilst producing the viral protein. This stimulates the production of antibodies. Under normal circumstances exposure to a viral disease would be countered (in vivo) at various levels enabling the body to steadily increase its immune response. By contrast, the injection of vaccines directly into the blood system overpowers the normal immune response leading to its rapid depletion. It is now suspected that long-term persistence of viruses and other proteins may produce chronic disease i.e. instead of producing a genuine immunity the vaccines are altering the body's systemic and biochemical stability, suppressing the production of differing types of white blood cells and hence immune function. Furthermore the introduction of many vaccines (up to 30 in a typical vaccination schedule) introduces a large number of foreign proteins which may be sufficient to ensure that immune function never returns to baseline and/or that immune biochemistry is fundamentally altered. Consequently there now exists a growing concern which links immunizations to the huge increase in recent decades of auto-immune diseases e.g., rheumatoid arthritis, multiple sclerosis, lupus erythematosus, lymphoma, leukemia, autoimmune demyelinative optic neuritis, diabetes mellitus, etc. …

The prevailing reaction conditions—the consequence of protein expression which has been influenced by previous vaccines—will also affect the introduction of each modified live virus. Each will depress immune function. The greater the number of viruses and foreign proteins (1) the greater the influence upon immune function and the time required for recovery from each vaccination; (2) the greater their influence upon DNA, gene and chromosome structure and function, the greater will be the risk of protein inhibition, system dysfunction, reproduction, etc. The greater the amount of vaccines, introduction of foreign proteins and hence of alterations to the body's biochemistry the greater the risk that the body's immune function no longer recognizes or responds to existing vaccines or diseases and/or that its immune response has been altered and/or that sugar chains attached to an antibody alters its ability to bind to its receptors. This may lead to mutated forms of disease e.g. the reemergence of whooping cough, and a differentiated disease profile e.g. up to 30 per cent of individuals with a persistent cough are infected with B. pertussis. Furthermore enhanced susceptibility to virus infection by vaccines is documented. This could enable tougher strains to flourish. …

That vaccines suppress natural immune function is not in dispute e.g. those with naturally low levels of immune function (immigrants from tropical climates) show greater predisposition to autistic spectrum disorders."[109]

"'If you take water into your system as drink, it goes into your bloodstream directly from the stomach. But if you take fats, they move into your lymphatic system. When you take other substances like carbohydrates and proteins, they go into the intestines, and from there are passed through the liver, as the body's chemical censor, before they go into the blood and the circulation throughout the body. Most of your vaccination serums are proteins, and are not censored by the liver. Consequently vaccinations can be a terrific shock to the system.' Injections of foreign substances—viruses, toxins, and foreign proteins—into the bloodstream, that is, vaccinations, have been associated with diseases and disorders of the blood, brain, nervous system, and skin. Rare diseases such as atypical measles and monkey fever as well as such well-known disorders as premature aging and allergies have been associated with vaccination. Also linked to immunizations are diseases such as cancer, leukemia, paralysis, multiple sclerosis, arthritis, SIDS, and most recently AIDS."[110]

Many other negative effects are associated with vaccination. For all of these diseases and conditions to be connected with vaccines might seem unbelievable, but there is much harm in placing compounds and biological substances into our bodies that were never meant to be injected. We put things that God specifically told His people to *avoid*, like disease, into the *"life of the flesh"* (Lev 17:11).

God warned His people what would happen to them if they did not obey His law. We are not the Israelites, and are not under the law. However, the law that God commanded, and that the Israelites agreed to keep, involved effective methods to prevent sickness, and to keep it from spreading. Whether these were supernatural consequences specifically for the Jews, or the natural consequences of living contrary to God's laws that affect all men, there is still something for Christians—"*...upon whom the ends of the world are come*" (1Co 10:11)—to learn (Rom 15:4) from this passage of Scripture.

Deuteronomy 28:58 **If thou wilt not observe to do all the words of this law that are written in this book, that thou mayest fear this glorious and fearful name, THE LORD THY GOD;** 59 **Then the LORD will make thy plagues wonderful, and the plagues of thy seed, even great plagues, and of long continuance, and sore sicknesses, and of long continuance.** 60 **Moreover he will bring upon thee all the diseases of Egypt, which thou wast afraid of; and they shall cleave unto thee.** 61 **Also every sickness, and every plague, which is not written in the book of this law, them will the LORD bring upon thee, until thou be destroyed.** 62 **And ye shall be left few in number, whereas ye were as the stars of heaven for multitude; because thou wouldest not obey the voice of the LORD thy God.**

The encephalitis, DNA damage, and altering of the immune system caused by vaccination culminate into a trifecta of disease, disfigurement, and death. Everything about the process of vaccination that causes these changes goes against the wisdom of God's Word, and we can see with our own eyes the deteriorating health of our nation.

I will mention a few more negative effects of vaccination, and while there is much evidence to support these claims, I will limit myself and only briefly comment on each issue.

Cerebral palsy:

"Cerebral palsy cases probably are equally insidious vaccine-related neurological problems. All countries with intensive or mandatory vaccination of infants experienced a 400% increase in the incidence of cerebral palsy in the last fifty years, not diagnosed at birth."[111]

Oculomotor palsy:

"Measles Mumps Rubella (MMR) vaccination is known to cause some serious adverse events, such as fever, rash, gland inflammation and neurologic disorders. These include third and sixth cranial nerve palsies. The case reported describes a partial recurrent oculomotor palsy associated with systemic symptoms following MMR vaccination in a healthy young child. The oculomotor palsy did not recover completely during the follow-up."[112]

Allergies, Asthma, Eczema:

"Vaccines, like any other noxious substances, do not immunise when injected into the bloodstream, rather they sensitise. This was known more than 100 years ago and was well described in medical journals like The Lancet. The upsurge of allergies in children in the past 50 years is the result of repeated multiple injection of the foreign antigens in vaccines."[113]

"...asthma, chronic bronchitis, eczemas, allergies and even cancer... The enormous increase in the incidence of these diseases, and the appearance of a great variety of immunodeficiencies plaguing humanity, and especially children, since the introduction of mass vaccination programmes, is well documented in medical literature."[114]

Learning Disabilities, etc.

"According to the testimony of the assistant secretary of health, Dr. Edward Brandt. Jr., before the U.S. Senate Committee on May 3, 1985, every year 35,000 children suffer neurological reactions because of this vaccine [DPT]. ...'some of the learning disabilities we see in children of school age may actually be signs of neurological damage suffered during a vaccine reaction."[115]

"'Of course, if a child develops encephalopathy, one part of the possible residual from that might be a learning disability.' Reports in the scientific literature and by parents suggest that pertussis vaccine damage covers the spectrum from death and severe retardation to such milder forms of neurological damage as learning disabilities. ... 'It would be out of keeping with other adverse reactions to other substances if the only cases were severe ones. More likely there are numerous cases of lesser damage and lesser disability.'"[116]

"The rise in illiteracy, semiliteracy, mental retardation, and criminality correlates with the increase in childhood vaccination programs."[117]

Activity-Limiting Chronic Conditions:

"Paralleling this increase in neurological damage is the increase in 'activity-limiting chronic conditions' such as childhood respiratory diseases and ear and eye diseases."[118]

Constant illness and infections:

"The mother cannot understand why her previously healthy child is suddenly 'never well.' The child may be sick for months with the mother taking him from doctor to doctor in search of a cure. Although never sick enough to be put in a hospital, the child never recovers his previous good health and continues to be plagued with constant 'colds,' ear infections, or the onset of allergies."[119]

"In the following weeks, Sean was transformed from a healthy, happy, and intellectually advanced child to one who was sick with constant diarrhea, ear infections, runny nose, and upper chest congestion. He lost his appetite and stopped growing. He became hypersensitive and tired easily without warning. His mother thought the abrupt personality and behavior change was due to his 'never being well anymore.' Doctors prescribed antibiotics and antihistamines, but nothing helped."[120]

Seizures:

"In Poland, researchers reported that 13 of 17 children who were given DT vaccinations showed significant changes in their electroencephalograms and the appearance of seizure activity for the first time or intensification of previously present seizure activity."[121]

Diabetes:

"An equally serious issue, about which less information has been developed, is the relationship between the vaccine and diabetes. Diabetes has been reported to be a complication of a very serious case of whooping cough. Four researchers in France, Western Germany, and Great Britain published an article in 1982 detailing the role the DPT vaccine played in causing diabetes in a sixteen-month-old girl who was genetically predisposed to diabetes and who suffered from a viral infection that attacked her pancreas. They concluded that 'an insulinotropic variant virus, genetic predisposition, and perhaps some uncontrolled adjuvant factors, e.g. steroid therapy and DPT vaccination, may have determined insular damage and anti-islet autoimmune reactions, leading to insulin-dependent diabetes mellitus."[122]

Cranial Nerve Damage

"The DPT vaccine is capable of causing all possible cranial nerve disorders: from blindness, deafness, and muteness to the less severe dyslexia, stuttering, or auditory processing deficit. ...sometimes these children have 'gaze palsy,' difficulty moving the eyeball from side to side or up and down. Or they will be crosseyed [sic]... How much of the blindness discovered in six-month-old children and diagnosed as 'congenital' is actually vaccine-induced? No one knows. By the time the blindness is diagnosed, the baby will have had two or three DPT shots—which are capable of causing blindness."[123]

Plenty of studies conclude there is no correlation between vaccination and any negative side effects whatsoever, but these are usually funded by entities with a vested interest in vaccination. Before considering a study reliable, it is necessary to "follow the money," and examine the parameters of the study, such as identifying the control group. Safety study results can be nullified by researchers "...disguising adverse reactions with bogus placebos"[124] in the same way other studies can be rendered invalid with a bogus control group. Privately funded research is unbiased, and these studies usually do not have outcomes favorable to vaccination.

"Private research groups have done many studies on vaccines, but the [capital] for such endeavors has always come out of their own pockets. These are the groups that repeatedly tell us vaccinated kids are among the sickest in the world."[125]

Health problems are so ubiquitous, even among children, it seems as though nearly all children are affected with some sort of ailment. Among unvaccinated children, however, these health problems are nearly nonexistent. There simply aren't as many unvaccinated kids, and those who make millions selling vaccines will suppress the fact that **unvaccinated children are healthier**.

"How is the health of unvaccinated children? Do they die of childhood diseases in droves? The answer is no, they don't. There is a general notion that unvaccinated children get sick more often. Indeed, it may be the opposite. ... Dr. Philip Incao...treated about 500 children in his Harlemville, New York practice, and a large number of them were not immunized. I inquired about the health of these unimmunized children and asked if he saw more or less incidence of asthma, allergies, hyperactivity, learning disabilities, and seizures among them. He said he saw much less—almost none..."[126]

Some scientists claim that vaccine damage due to fetal DNA contamination is "unsupported and implausible," due in part to this "fact":

"The immune system has evolved to not recognise self-nucleic acids (DNA and RNA) as this would provoke an autoimmune response."[127]

So, because this DNA would cause an autoimmune response, therefore the body has evolved to prevent this? I'm not sure how anyone could know this, as macroevolution has never been observed, and we've seen that the immune system is only superficially understood. While this person is highly intelligent, much more so than myself, God's word gives us wisdom that is greater than worldly wisdom. We have been told the opposite, but it is not strange for those with a Biblical worldview to accept that harm can come from injecting the DNA of murdered children into our living children; it is an abomination.

It also should not be far-fetched to think that changing man's immune system would cause a denigration of its integrity. Man's designs cannot improve upon our God-given immune systems; they can damage the immune system at a critical point in its development. God's understanding is infinite, and God is good. We are unwise to go against or change His designs. It is certainly true that a healthy person does not need a physician (Luk 5:31); but Christ also said those that are sick **do** need a physician. A perfectly developed human being does not need any type of aborted fetal medicine to maintain that state of health. Of course, there are situations where medical intervention is necessary; aborted fetal cell line vaccines do not fall under this category. The goal of this book is to show that health will never come from any "medicine" made from an aborted fetal cell line; the changes these products make in the human body will not better it. There are obviously children born with defects that have their lives greatly improved, or saved, because of medical interventions and surgery: these situations are not the subject of this book.

Man cannot do anything better than God. We can take things from God's creation, work *with* His designs rather than against, and make better medicines, such as intravenous vitamin C (see Pro 8:12); but we cannot make our bodies better than God made them. Yes, we live in a fallen world, and things have changed since Adam; that does not mean the only way to be healthy is to use aborted fetal cell lines. Simply because the world has fallen does not mean we need to fall along with it, and allow some evil to supposedly accomplish good: *"Because it is written, Be ye holy; for I am holy"* (1Pe 1:16). Trying to change man's immune system will only detract from the way God made it.

Man also cannot create genuine immunity; we cannot do in a lab what God does in nature. Some vaccines are certainly effective for a period of time, and the *incidence* of some diseases has reduced since their mass implementation. However, the *mortality* of these diseases was greatly reduced *before* the widespread use of vaccines. Diseases that can be deadly in squalid living conditions become mild illnesses in our modern world. While the burden of disease is not just measured in mortality, the undeniable burdens of vaccinations, such as SIDS and autism, are much worse than those of mild childhood illnesses today.

"The incidence of vaccine side-effects may now be sufficiently great to question the claims that the risks from the disease exceed that of vaccines."[128]

"Calculations based on the mortality of whooping-cough before 1957 predict accurately the subsequent decline and the present low mortality. Notifications of incidence, though variable and incomplete, follow the same pattern of steady decline in the United Kingdom and are unaffected either by small-scale vaccination beginning about 1948 or by nationwide vaccination beginning in 1957. ... No protection by vaccination is demonstrable in infants. Adverse reactions and neurotoxicity following vaccinations were studied in 160 cases. In 79, the relationship to pertussis vaccine was strong. In 14 of these cases, reaction was transient but characteristic of a syndrome of shock and cerebral disturbance, which, in the other 65 cases, was followed by convulsions, hyperkinesis, and severe mental defect. It seems likely that most adverse reactions are unreported and that many are overlooked. Precise information about the efficacy and safety of this vaccine is lacking, because existing provisions, national and international, for epidemiological surveillance and evaluation are inadequate. The claim by official bodies that the risks of whooping-cough exceed those of vaccination is questionable, at least in the U.K."[129]

"There are no randomized controlled studies comparing the total health outcomes of the vaccinated versus the unvaccinated. Scientists do not know the cumulative impact of the childhood vaccination schedule. Vaccines are approved and licensed individually, yet the CDC recommends that doctors give multiple vaccines at once. ... Research confirms that fewer than 10 percent of doctors report adverse vaccine events."[130]

With all of these side effects, doesn't it seem like vaccine manufacturers would face so many liability claims they would go bankrupt? That is exactly what was about to happen in the 80s, until the government stepped in and made it illegal to sue vaccine manufacturers. That's right, those who profit from vaccines are free from any liability that might come from permanently disabling or killing your child. A vaccine manufacturer "...can't be taking the same risks as the recipients of its vaccine..."[131] If vaccines were truly life-saving products with "few adverse effects," as they are lauded to be, why did they need a government bailout law—The National

Childhood Vaccine Injury Act of 1986—to keep them from facing mountainous litigation that surely would've bankrupted them?

The National Immunization Compensation Program was also established, with the intent to quickly, generously, and efficiently compensate families of vaccine-injured children; what it actually became is much different. The sad truth is the pharmaceutical industry has "…remarkable liability protections…" and "…as a result of the act, the vaccine-injured and their families face unprecedented hurdles to obtain compensation."[132]

"'The deck is stacked against families in Vaccine Court. Government attorneys defend a government program, using government funded science, before government judges.' … 'I worked with Congress in the early 1980s on that (vaccine compensation) law and have watched it be turned into a cruel joke as 2 out of 3 vaccine injured children are denied federal compensation for their often catastrophic vaccine injuries because HHS (Department of Health and Human Services) and the Department of Justice officials fight every claim, viewing every award to a vaccine injured child as admission that vaccines can and do cause harm.'"[133]

"…the unknown numbers of vaccine-damaged children and their parents who, like soldiers in a terrible war, have given their lives for a society that is reluctant to accept responsibility for the sacrifice it has required them to make."[134]

"The 1986 Act granted unprecedented, economic immunity to pharmaceutical companies for injuries caused by their products and eviscerated economic incentive for them to manufacture safe vaccine products or improve the safety of existing vaccine products. Congress therefore charged the Secretary of HHS with the explicit responsibility to assure vaccine safety."[135]

According to The National Childhood Vaccine Injury Act of 1986, the United States Department of Health and Human Services (HHS) is required to submit a biennial report showing the improvements in vaccine safety attained over the previous two years. The Informed Consent Action Network (ICAN) filed a Freedom of Information Act requesting these reports in 2017, but were ignored. Here is what happened after eight months of the government failing to provide this information.

"ICAN was therefore forced to file a lawsuit to force HHS to either provide copies of its biennial vaccine safety reports to Congress or admit it never filed these reports. The result of the lawsuit is that HHS had to finally and shockingly admit that it never, not even once, submitted a single biennial report to Congress detailing the improvements in vaccine safety. This speaks volumes to the seriousness by which vaccine safety is treated at HHS and heightens the concern that HHS doesn't have a clue as to the actual safety profile of the now 29 doses, and growing, of vaccines given by one year of age [in 2018]. … If HHS is not, as confirmed in [c]ourt this week, even fulfilling the simple task of filing a biennial report on vaccine safety improvements, there is little hope that HHS is actually tackling the much harder job of actually improving vaccine safety."[136]

"The vaccine schedule of the early 1980s was causing so many injuries and lawsuits that manufacturers threatened to stop making vaccines due an alleged lack of profitability. … The 1983 immunization schedule for children up to age of six recommended ten injections of three vaccines. … The CDC 2015 immunization schedule for children by the age of six recommends thirty-seven injections of ten vaccines."[137]

HHS has not completed **one single report** showing an increase in vaccine safety in the last 30 years, which they are required by law to do every two years. If the government-mandated immunization schedule was causing so many injuries in the 1980s, can the current schedule, which is *much more aggressive* and has more than tripled the vaccine doses, be considered safe? This act was passed because of liability claims from vaccine-induced injuries and not one shred of evidence exists showing that anything has been done to make these products safer for our children. Vaccines are legally considered to be "unavoidably unsafe" products, and there have been no improvements in safety (aside from replacing DPT with DTaP) in the last 30 years, yet many additional unsafe vaccines *have been added* to the schedule. Is this good enough for your children? Have those in charge of vaccine safety earned your trust?

Why have most of us never heard any of this information? Like the fact that there certainly are links to spontaneous abortion, SIDS, autoimmune disease, and autism? The fact that vaccines have not caused diseases to disappear and that they are responsible for terrible side effects, makes the vaccines worse than the diseases they are supposed to protect against. Why have you never been told these things? Because the "*…love of money is the root of all evil…*" (1Ti 6:10). Again, it is very profitable to make every healthy person believe they need to buy something to keep them healthy. Doubts are not "allowed to exist"; it is of the utmost importance that trust in immunization is maintained. The FDA also made these statements in that same publication:

"The agency has concluded, however, that this requirement is unnecessary to assure the safety, purity, and potency of the oral poliovirus vaccine used in clinical trials and could be the cause of a meaningless waste of effort and vaccine by a manufacturer conducting clinical studies in the United States or abroad. The manufacture of a viral vaccine is a complex operation involving living organisms. Therefore, it is inevitable that occasionally an attempt to manufacture a safe, pure, and potent oral poliovirus vaccine will be unsuccessful [despite] the use of good manufacturing practices. ...FDA has concluded that it is in the best interest of the public health to amend § 630.11 to eliminate the unnecessary requirement that the vaccine used in clinical trials show satisfactory results in all tests applicable to lots used in clinical trials, and thus avert any possible loss of confidence in the polio immunization program."[138]

Many of the findings from recent studies remain unheard because: "It takes an estimated average of seventeen years before evidence from scientific research is incorporated into day-to-day clinical practice."[139] It is also tantamount to career suicide for a doctor to question vaccines, and most doctors have no idea that there is anything to question in the first place.

"It is difficult to get a man to understand something when his salary depends upon his not understanding it."—Upton Sinclair

"Historically, doctors who have dared to change things significantly have been ostracized and have had to sacrifice their careers in order to hold to their ideas. Few doctors are willing to do either. ... It's well-known that Modern Medicine doesn't act on scientific knowledge until public awareness grows strong enough to *demand* it."[140]

Robert Mendelsohn, MD, quoted above, also suggested becoming as informed as possible, so that you can be "...on an equal—or better—informational footing..."[141] as your, or your child's doctor. He also suggests patients to be highly skeptical of medical professionals, and not to take any doctor's advice unless it has absolutely been proven to be true. (If you think that's harsh, you should read what he actually said.[142])

"When the function of academic research is to foster and sustain a delicate fabric of collaboration, no one will bite any hand that feeds them—particularly in the climate that exists today. 'To serve the public interest, government advisory committees must be independent of industry, but such committees cannot be relevant and effective if isolated from the expertise and experience of the industry, which is the principal funder of vaccine research and development.' Until this political triangle is broken, parents must know that the health of their children rests upon their own research and good judgment."[143]

"Because the whole truth isn't told, adults are the only line of defense for themselves. Until the minds of pediatricians are emancipated, parents will remain the best line of defense for their children."[144]

"How can there be no outcry against rising infant mortality and skyrocketing childhood diagnoses for attention deficit disorder, autism, learning disabilities, and asthma?"[145]

With all that you have just read, is it wise to blindly follow the advice of someone who benefits financially from vaccines? Or who benefits financially from the many conditions that are associated with vaccine use? I truly believe that most doctors want what is best for their patients, but that doesn't mean we cannot compare what they say to what God has said in His Word. The sad fact is: "...most physicians know little about vaccines beyond what they are told by the pharmaceutical sales representatives who sell them."[146]

1 Timothy 6:10 **For the love of money is the root of all evil: which while some coveted after, they have erred from the faith, and pierced themselves through with many sorrows.**

Proverbs 14:12 **There is a way which seemeth right unto a man, but the end thereof are the ways of death.**

"What if it were true that the way we now vaccinate children causes more death, chronic disease, and disability than it prevents in America? Blanket assurances that this can't happen are naive. Pharma-funded studies to diffuse concern—whether the money flows first through university foundations, hospitals, or nonprofit—are obviously flawed, inherently biased, and can't yield clear information."[147]

"Any pediatrician, doctor, or state or federal public health official who tells you that vaccines are completely safe, that adverse reactions to vaccines don't exist, or that vaccine-induced injuries are so rare that they virtually never occur is either ignorant or is committing scientific fraud."[148]

"As a concerned pediatrician, I can arrive at only one conclusion. Unvaccinated children have by far the best chance of enjoying marvelous health. Any vaccination at all works to cripple the chances of this end."[149]

Chapter 5

Vaccinated vs. Unvaccinated Pilot Study

In 2017, there was a pilot study comparing vaccinated and unvaccinated homeschooled children. The study states:

"Vaccinations have prevented millions of infectious illnesses, hospitalizations and deaths among U.S. children, yet the long-term health outcomes of the vaccination schedule remain uncertain. Studies have been recommended by the U.S. Institute of Medicine to address this question. This study aimed 1) to compare vaccinated and unvaccinated children on a broad range of health outcomes, and 2) to determine whether an association found between vaccination and neurodevelopmental disorders (NDD), if any, remained significant after adjustment for other measured factors… Although short-term immunologic and safety testing is performed on vaccines prior to their approval by the U.S. Food and Drug Administration, the long-term effects of individual vaccines and of the vaccination program itself remain unknown… The safe levels and long-term effects of vaccine ingredients such as adjuvants and preservatives are also unknown."[1]

Here are some of the most interesting findings from this study.

"With regard to acute and chronic conditions, vaccinated children were significantly less likely than the unvaccinated to have had chickenpox and pertussis but, contrary to expectation, were significantly more likely to have been diagnosed with otitis media, pneumonia, allergic rhinitis, eczema, and NDD. The vaccinated were also more likely to have used antibiotics, allergy and fever medications; to have been fitted with ventilation ear tubes; visited a doctor for a health issue in the previous year, and been hospitalized. The reason for hospitalization and the age of the child at the time were not determined, but the latter finding appears consistent with a study of 38,801 reports to the VAERS of infants who were hospitalized or had died after receiving vaccinations. The study reported a linear relationship between the number of vaccine doses administered at one time and the rate of hospitalization and death; moreover, the younger the infant at the time of vaccination, the higher was the rate of hospitalization and death. The hospitalization rate increased from 11% for 2 vaccine doses to 23.5% for 8 doses, while the case fatality rate increased significantly from 3.6% for those receiving from 1-4 doses to 5.4 % for those receiving from 5-8 doses. …

Table 7 shows that the factors associated with NDD in unadjusted logistic regression analyses were: vaccination; male gender; adverse environment, defined as living within 1-2 miles of a furniture manufacturing factory, hazardous waste site, or lumber processing factory; maternal use of antibiotics during pregnancy; and preterm birth. Two factors that almost reached statistical significance were vaccination during pregnancy and three or more fetal ultrasounds. …

After adjustment for all other significant factors, those that remained significantly associated with NDD were: vaccination; male gender; and preterm birth. The apparently strong association between both vaccination and preterm birth and NDD suggested the possibility of an interaction between these factors. In a final adjusted model designed to test for this possibility, controlling for the interaction of preterm birth and vaccination, the following factors remained significantly associated with NDD: vaccination, nonwhite race and male gender. Preterm birth itself, however, was not significantly associated with NDD, [but] the combination (interaction) of preterm birth and vaccination was associated with 6.6-fold increased odds of NDD. …

Vaccination was strongly associated with both otitis media and pneumonia, which are among the most common complications of measles infection. The odds of otitis media were almost fourfold higher among the vaccinated and the odds of myringotomy with tube placement were eight-fold higher than those of unvaccinated children. …

No significant differences were seen with regard to hepatitis A or B, high fever in the past 6 months, measles, mumps, meningitis (viral or bacterial), influenza, or rotavirus."[2]

(According to this study's sister study which specifically looked at preterm birth, vaccination and neurodevelopmental disorders, a preterm, fully-vaccinated infant is **14.5 times** more likely to be developmentally disabled versus a full-term, unvaccinated child.[3] However, without vaccination this association disappears.)

In this study, both vaccinated and unvaccinated groups experienced "vaccine-preventable" diseases—measles, hepatitis A and B, mumps, meningitis, and influenza—at the same rate. However, the vaccinated were far more likely to experience ear infections, and to have tubes placed than the unvaccinated. These children did not have any additional protection from measles versus unvaccinated children but still experienced complications associated with measles. The study does not list measles occurrences, just the above statement, so if none of the children had measles there wouldn't be any differences between the groups. In this small study of just under 700 children, this doesn't speak to the vaccine being ineffective as some studies would suggest, but would make it seem as though the vaccine is not as necessary as we are told, especially in light of the many questions regarding its safety.

This group of children was maybe too small to get a good measure of the incidence of some of these infectious diseases, but we have seen these diseases still occur in highly vaccinated populations that we are told should have herd immunity. Also paradoxically, older groups of people who do not get all of their booster shots and have coverage numbers considered too low for herd immunity do not experience outbreaks that only affect these "under" vaccinated populations.

Parents, if vaccinated versus unvaccinated children experience around the same amount of febrile infections—aside from chickenpox and pertussis in this small study—why take the route associated with chronic health conditions? Regarding whooping cough, if a pediatrician has given a child a vaccination for pertussis, they are not likely to look, test for, and diagnose pertussis. Many studies prove this vaccine is not very effective, and other studies demonstrate gross underreporting of pertussis:

"Our study suggests that there is substantial underreporting of pertussis, that severe complications of pertussis (including hospitalizations) are reported preferentially to the CDC, and that the national health impact of pertussis based on these indicators is considerably higher than previously published reports have suggested."[4]

"The general and chronic under-reporting of infectious diseases after the introduction of vaccines of any kind in the US is quite well-known and often criticised."[5]

Profits are not only accrued from the sale of vaccines, the negative health effects following vaccination require treatments which would also financially benefit medical establishments and the pharmaceutical industry. Vaccinated children are more likely to take medications, get drainage tubes placed in their ears, and be hospitalized. Now multiply these expenses by every child in the country. Doesn't this look a little bit like a "create a problem and sell the cure" situation?

The study concludes:

"Nevertheless, the study findings should be interpreted with caution. First, additional research is needed to replicate the findings in studies with larger samples and stronger research designs. Second, subject to replication, potentially detrimental factors associated with the vaccination schedule should be identified and addressed and underlying mechanisms better understood. Such studies are essential in order to optimize the impact of vaccination [on] children's health."[6]

Ask yourself why the government has not performed a study like this one. If we as a nation want to be certain the vaccine schedule is safe, why can we not compare unvaccinated children to their vaccinated counterparts? Over ten years ago, a bill was introduced called the Comprehensive Comparative Study of Vaccinated and Unvaccinated Populations Act of 2007, (H.R.2832) which:

"Requires the Secretary of Health and Human Services, acting through the Director of the National Institutes of Health (NIH) to conduct a comprehensive study to: (1) compare total health outcomes, including the risk of autism, between vaccinated and unvaccinated U.S. populations; and (2) determine whether vaccines or vaccine components play a role in the development of autism spectrum or other neurological conditions. [Also] Requires the Secretary to seek to include in the study U.S. populations that have traditionally remained unvaccinated for religious or other reasons."[7]

This bill never became a law. Why have these investigative efforts been continuously thwarted by our government and its "health" agencies?

"These studies should be funded by the government but conducted by independent researchers who have no stake in a particular outcome. The studies cannot be conducted by agencies that promote vaccination and have a mission to ensure public confidence in vaccines. The government's failure to fund the basic scientific studies of these medical products while simultaneously recommending or mandating them raises profound ethical questions."[8]

A growing number of parents are not vaccinating according to the CDC's schedule. This movement is not going away; the truth is gaining traction. Enough children are available to do a larger study, and to do multiple studies. We don't know if vaccines are associated with other chronic diseases or our top killers because no one will compare vaccinated children to *completely* unvaccinated children. Until more parents become aware, this ongoing nationwide experiment is just going to continue.

"In the meantime, entire countries' populations are unwitting gamblers in a game that many might very well choose not to play if they were given all the rules in advance. ... Millions of children are partaking in an enormous, crude experiment, and no sincere, organized effort is being made to track the negative side effects or to determine the long-term consequences. Since long-term studies on the adverse effects of vaccines are virtually non-existent, their widespread use in the absence of informed consent and adequate safety testing constitutes medical experimentation. As the American Association of Physicians and Surgeons and the National Vaccine Information Center have pointed out, this is a violation of the first principle of the Nuremberg Code, 'the centerpiece of modern bioethics.'"[9]

I would also like to clear the misunderstanding that unvaccinated children have to be homeschooled. All states do have immunization requirements, but there are also exemptions.

"All 50 states have legislation requiring specified vaccines for students. Although exemptions vary from state to state, all school immunization laws grant exemptions to children for medical reasons. There are 45 states and Washington D.C. that grant religious exemptions for people who have religious objections to immunizations. Currently, 15 states allow philosophical exemptions for those who object to immunizations because of personal, moral or other beliefs."[10]

In most states, unvaccinated children do not have to be homeschooled. It is actually just a matter of writing a letter to the proper entity, stating which vaccines are against your religious beliefs. Simply look online for a template or sample letters you can use. If the group or individual requiring your vaccination doesn't realize that not accepting your exemption would make them liable for anything that should happen, kindly make them aware of their error.

"The people never give up their liberties but under some delusion."—Edmund Burke

All of this information is important because we want the public to be aware and to not consent to having medical freedoms taken away from us, as people like Paul Offit would like. Offit has stated:

"...babies could theoretically respond to a hundred thousand vaccines at one time. Of course, we're not saying that babies should get a hundred thousand vaccines at once. We're only saying that they could handle it. Indeed, given that babies are constantly confronted with trillions of bacteria and that each bacterium contains thousands of immunological components, this shouldn't be surprising. In a sense, babies are responding to such an assault every day."[11]

Constantly interacting with bacteria does not make a baby capable of handling a vaccine being injected into his or her body; this bypasses the mechanisms God designed to handle these bacterial interactions. While Offit's statement sounds theoretically plausible, it is very unrealistic and unscientific. Just like the snake venom: our digestive system can handle it, our circulatory system cannot. **A baby receiving a vaccine and a baby being exposed to bacteria are not the same thing.** Babies do not get injected with every bacterium they encounter along with aborted fetal DNA, immune adjuvants, and harmful substances like formaldehyde. Babies are not exposed to such an assault every day, just on the days they visit their doctor. Also, why is it that babies can supposedly handle tens of thousands of vaccines but not a case of chickenpox?

"Vaccination is the epitome of ignorance and the unscientific approach to illness. Yet, modern medicine considered itself scientific. Nothing could be further from the truth."[12]

If Offit gets his dearest desire of taking away our freedom to reject unwanted vaccines, I wonder if he and his family will be exempt from that law? If, and I stress, *if* his children have received any vaccines, I wonder how many they have received at one time? I wonder if his own children received the rotavirus vaccine, of which patent he owns. Medical authorities claiming children could handle thousands of vaccines, simply because they can handle thousands of natural bacteria interactions every day, should alarm us. If our exemptions are taken away, how many vaccines will be forced upon our children? If vaccines were so great, they need not be legislated.

"The truth is that if a product is overwhelmingly safe and effective, it will not need legislation for mass acceptance. And conversely, if it proves to be questionable, legislation is the only way to insure mass acceptance."[13]

It is a shame there are states which do not allow religious exemptions to those who do not want aborted fetal DNA injected into their bodies. It **should be a crime** to force parents to use a medical intervention that can result in the deaths of susceptible children.

When the religion of the state is evolution, I suppose "Only The Fittest Shall Survive" is the first and greatest commandment.

"How can anyone justify a public policy that forces vulnerable infants and young children to bear the burden of risk and suffering—including neurological damage, seizures, permanent disability, and death—for 'herd immunity'?"[14]

Again, notice the parallel between smallpox vaccination and modern vaccination:

"While thus utterly powerless for good, vaccination is a certain cause of disease and death in many cases, and is the probable cause of about 10,000 deaths annually by five inoculable diseases of the most terrible and disgusting character, which have been increased to this extent, steadily, year by year, since vaccination has been enforced by penal laws! ... If these facts are true, or anything near the truth, the enforcement of vaccination by fine and imprisonment of unwilling parents, is a cruel and criminal despotism, which it [behooves] all true friends of humanity to denounce and oppose at every opportunity. Such legislation, involving as it does, our health, our liberty, and our very lives, is too serious a matter to be allowed to depend on the misstatements of interested officials or the dogmas of a professional clique."[15]

"The public will eventually wake up and see forced vaccination for what it is—a dangerous, money-grabbing, scientifically invalid fraud."[16]

Some of the founding fathers tried to establish this country with medical freedoms equal to that of our religious freedoms, recognized as being inalienable rights endowed by the Creator. Those who work in health care and government should have the right to refuse medical interventions they do not want. An excerpt from a speech made by Dr. Benjamin Rush, a congressman who signed the Declaration of Independence, with added parenthetical statements as recorded in Eleanor McBean's book *The Poisoned Needle,* follows:

"The Constitution of the Republic should make provision for medical freedom as well as for religious freedom. To restrict the art of healing to one class of men (such as medical doctors) and deny equal privileges to others (also trained in healing) will constitute the Bastile [sic] of medical science. All such laws are unAmerican and despotic. They are fragments of monarchy and have no place in a Republic."[17]

To force citizens to take a medication they do not want is completely against the Biblical principles on which this country was originally founded. While religious freedom is certainly more important, it is a shame that medical freedom was not won as well. We must wonder, why was this prevented?

2 Corinthians 3:17 **Now the Lord is that Spirit: and where the Spirit of the Lord is, there is liberty.**

Since researching this subject for the last couple years, I've seen information control regarding vaccines because this information supposedly "harms children." This statement is a dishonest ploy to keep decent people ignorant. Anytime someone controls information, or there is no *liberty—something is very wrong.* Most medical professionals truly believe that vaccines have saved humanity. They live in the same "Truman Show" setup of vaccination dogma that everybody else does, and on top of that have received an education which affirms the common misconceptions about the benefits of vaccination. Those in charge of our media and medicine *know better.* When one understands that vaccines did not cause diseases to disappear but instead have seriously injured an unknown number of children, the truth becomes clear: the idea that our information must be controlled shows that mainstream misinformation harms children by keeping people ignorant.

I'm very thankful for religious freedom. I'm thankful for this country, and everything that God has placed in my life. Compared to any other time and place that God could have put me, I'm very thankful to be here, and to have many freedoms that are denied to others. God has placed me where He willed, so that I could come to a saving knowledge of Him, as He wills for all men. Pray for the leaders of this country; our wickedness is bringing about our destruction. I'm certainly ready for the rapture, and for Jesus to rule and reign with *"a sceptre of righteousness"* (Heb 1:8); but until then, I want to live a quiet and peaceable life.

Psalm 107:33 **He turneth rivers into a wilderness, and the watersprings into dry ground;** 34 **A fruitful land into barrenness, for the wickedness of them that dwell therein.**

1 Timothy 2:1 **I exhort therefore, that, first of all, supplications, prayers, intercessions, and giving of thanks, be made for all men;** 2 **For kings, and for all that are in authority; that we may lead a quiet and peaceable life in all godliness and honesty.** 3 **For this is good and**

acceptable in the sight of God our Saviour; 4 **Who will have all men to be saved, and to come unto the knowledge of the truth.**

Chapter 6

Wait and Pray

Jeremiah 14:22 **Are there any among the vanities of the Gentiles that can cause rain? or can the heavens give showers? art not thou he, O LORD our God? therefore we will wait upon thee: for thou hast made all these things.**

Lamentations 3:25 **The LORD is good unto them that wait for him, to the soul that seeketh him.**

As a Bible-believing Christian, I also believe in individual soul liberty. You have the absolute right to choose what you will do. I fully respect that right; but I want to ask parents everywhere to wait and pray. Wait until you are adequately informed on this issue and pray for God's wisdom while you go through this process. If you feel your child must receive some ethically-made vaccines, that is your business, and your choice. I've mentioned Proverbs 18:13 a few times: if we answer a matter before we hear it, it is folly and shame to us. This same principle can be used to say: don't consent to a medical intervention without knowing all of the facts.

Proverbs 15:28a **The heart of the righteous studieth to answer**...

The CDC recommends all children receive a hepatitis B injection the day they are born. Many parents are opting to skip this vaccine. The only reason this vaccine is given to newborns is because they are more "accessible"[1] than adults, not because they are at risk of developing this disease as children. (If the mother is hepatitis B positive that is a different situation, about which I am admittingly uninformed. Please consult a physician who completely understands the risks of this vaccine.) Another reason infants are given this shot is because of the cost: "Cost of vaccinating infants is less than vaccinating adults, since a smaller dose is required."[2] Money should not be a motivating factor when deciding to give an "unavoidably unsafe" product to a **newborn infant**. "One could justifiably ask: is the system afraid to mandate vaccination for people who can talk and describe adverse reactions to the vaccines?"[3]

"Did you know that the groups at high risk for contracting hepatitis B are prostitutes, sexually active homosexual men, intravenous drug users, institutionalized children and adults, some health care workers, the military and people on hemodialysis? Hardly newborn infants! But then babies are 'accessible'!"[4]

I would like to share a couple studies with you, showing the waning immunity of the hepatitis B vaccine.

"The hepatitis B vaccine has been introduced for more than 3 decades. In Hong Kong, excellent vaccine coverage through an efficient public health care system, together with supplemental programmes and easy availability of the vaccine, meant that most young pregnant women, and university students at entrance, should have been protected. Yet significant correlations in the prevalence of HBV infection with age were found in these groups of subjects, increasing from low to high endemicity rates from late teenage to the early twenties. This can only be attributed to vaccine failure, and there is cumulating evidence that several factors are involved, including the failure to respond to a primary series of hepatitis B vaccination in infancy… The duration of protection conferred by hepatitis B vaccination in infancy should be re-examined and remedial measures undertaken if its long term protection is found to be insufficient. Otherwise, the efforts to control HBV infection, especially in high endemicity regions, with universal vaccination in infancy would be rendered futile. … Indeed, among young mothers born in Hong Kong who should have been protected by hepatitis B vaccination administered in infancy, HBV infection increased… Although few in number, these studies nevertheless suggested the increasing prevalence of HBV infection with age is likely to be related in part to vaccine failure in immunized subjects. … Full vaccination in infancy cannot guarantee complete protection, as up to 31.9% of neonatal vaccinees responded weakly thus rendering them susceptible to infection after childhood. … The concept of vaccine expiry, including the loss of the anamnestic response, following a full course of hepatitis B vaccination in infancy represents a paradigm shift."[5]

"Fifteen years after neonatal immunization with plasma-derived HB vaccine, a large proportion of children exhibited waning immunity. This poses the risk of breakthrough infection. A single booster augmented the serological response to the vaccine in most but not all subjects. In conclusion, our findings suggest that one or more booster immunizations are needed in seronegative subjects by at least 15 years following neonatal immunization with plasma-derived HB vaccine."[6]

In 2018, the year my daughter was born, her first pediatrician asked why our daughter had not received the hepatitis B vaccine. I didn't know much at the time, but had learned that there was absolutely no reason for her to get this vaccine. I asked the doctor if the vaccine would still provide immunity into adulthood, understanding there are many jobs that require employees to receive this vaccine, and we were told that it would. This last study quoted was *from 2004*. I've already mentioned that it takes an average of seventeen years before new medical scientific information goes from being realized to being implemented. When one sees how ignorant a practicing doctor can be in regard to information that could seriously affect their child's health, it goes to show that we must do our own research. It is wise to enlist knowledgeable professionals to aid us, but we cannot outsource our duties as parents.

Please understand that your child's doctor could very well be 20 years behind the newest medical research. This is more the fault of the system than the doctors; many times they are victims as much as anyone else. Whatever the cause, parents must inform themselves to protect their children. I agree with one doctor who wrote about his peers: "...I have known most medical doctors to be very hard-working and dedicated individuals. However, I believe that the bulk of medical practitioners are overworked, under-paid (thanks to managed care), highly stressed and [misguided]."[7] He also said that many doctors are so busy saving lives that they don't have time to find out why their efforts are necessary in the first place. That is why parents must research for the sake of their children. "*Prove all things...*" (1Th 5:21) is not an easy task in this day and age, but parents must rise to this challenge for their child's safety.

If you search the government's VAERS (Vaccine Adverse Event Reporting System) database for hepatitis B vaccine (Engerix-B,

Recombivax HB, and Heplisav-B) deaths (there are ***multiple death categories***, I only searched one) from **1 dose** for all of 2019, the results show there were 22 deaths reported; 8 of these deaths (36 percent) were among children under six months of age. Considering the total US population, this might not seem high if you believe this vaccine saves lives—if you were the parent of one of those children, these "low" numbers would be of no comfort to your shattered world. However, when one realizes how few adverse events are actually reported, these numbers might only be the tip of the iceberg. A government grant funded project performed by Harvard Pilgrim Health Care, Inc., explains:

"Adverse events from drugs and vaccines are common, but underreported. Although 25% of ambulatory patients experience an adverse drug event, less than 0.3% of all adverse drug events and 1-13% of serious events are reported to the Food and Drug Administration (FDA). Likewise, fewer than 1% of vaccine adverse events are reported. Low reporting rates preclude or slow the identification of "problem" drugs and vaccines that endanger public health. New surveillance methods for drug and vaccine adverse effects are needed. ... Unfortunately, there was never an opportunity to perform system performance assessments because the necessary CDC contacts were no longer available and the CDC consultants responsible for receiving data were no longer responsive to our multiple requests to proceed with testing and evaluation."[8]

A system that only accomplishes 1% of what it's designed to do, and this is how we are supposed to find out if there's a problem with the many vaccines given to nearly every child in the US? This is what happens when you put the government in charge of your children's health. There have also been more **documented** adverse reactions to the hepatitis B vaccine than there have been children with hepatitis B. Only God knows the true number of children harmed by the vaccine as only a small fraction of reactions are reported.

"According to a recent federal government study, serious adverse events after the vaccine - including 48 deaths - are reported three times as frequently as cases of hepatitis B in children under the age of 14."[9]

Another reason to seriously consider declining this vaccine is the amount of aluminum it contains: "On Day 1 of life, infants receive 17 times more aluminum than would be allowed if doses were adjusted per body weight."[10] According to the FDA: "The amount of aluminum in the recommended individual dose of a biological product shall not exceed: (1) 0.85 milligrams if determined by assay…"[11] This calculation, however, is based on vaccine efficacy requirements—*safety* is not a consideration.

"The Center for Biologics Evaluation and Research (CBER) sets the amount of aluminum per dose in biological products, including vaccines, to 850μg aluminum if measured by assay [.85 milligrams=850 micrograms=850μg]. Two additional levels are specified by the regulations (1140 and 1250μg respectively), depending on how the level is measured. … The dosing of aluminum in vaccines is based on the production of antibody titers, not safety science. … The 850μg of aluminum per vaccine FDA amount was derived from data that demonstrated that this amount of aluminum per dose enhanced the antigenicity and effectiveness of the vaccine, but does not include safety considerations. Current amounts of aluminum are not adjusted to body weight of an infant."[12]

Accordingly, the amount of aluminum allowed in parenteral drug products used in "total parenteral nutrition" is *much* lower.

"The aluminum content of large volume parenteral (LVP) drug products used in total parenteral nutrition (TPN) therapy must not exceed 25 micrograms per liter (microg/L)."[13]

Why is the aluminum limit **25 micrograms** for LVP drug products that can be injected, but **850 micrograms** for biological products, such as vaccines, that are also injected?

According to The Children's Hospital of Philadelphia, a pediatric hepatitis B vaccine that a newborn would receive contains 0.225 mg/dose[14] of aluminum; this measurement is in milligrams, not micrograms. A quick conversion shows this vaccine given to a newborn child the very day they are born contains 225 micrograms of aluminum, which is much lower than the limit for biological products, but is almost ten times the amount the FDA allows in LVP products. **Is it really worth injecting newborn babies with a neurotoxic substance for a disease that poses virtually no threat to them?** Furthermore, the FDA also states:

"Research indicates that patients with impaired kidney function, including premature neonates, who receive parenteral levels of aluminum at greater than 4 to 5 (micro)g/kg/day accumulate aluminum at levels associated with central nervous system and bone toxicity. Tissue loading may occur at even lower rates of administration."[15]

I don't understand why a premature baby can somehow handle the 225 micrograms of aluminum in the hepatitis B vaccine when much smaller amounts are associated with toxicity. Parents, unless you intervene, your newborn child will receive this vaccine. This next comment was made in regard to mercury, which has been removed from pediatric vaccines, except for "trace" amounts in certain shots; when looking at the large amount of aluminum presently in these vaccines, the principle still applies.

"A single vaccine given to a 6 pound newborn is the same as giving a 180-lb adult 30 vaccines on the same day."[16]

Aluminum's long-term effects on the immune system are also largely unknown. In Chapter 4, a study excerpt stated aluminum persists biologically in the immune cells of some individuals. One author surmised if the body signals for help and gets immune cells compromised with aluminum this could make certain situations worse. Why take this risk with our newborn children's immune systems that will largely determine their health for the rest of their lives?

At birth, a child's myelination process is just beginning—a development that takes *years*. Yet another reason to seriously consider declining, or at least postponing, the routine vaccinations your infant would otherwise receive.

"The one common denominator of developmental disabilities is nearly always related to demyelination of the cranial nerve fibers. Either impairment of the process of myelination or destruction of myelin results in cranial nerve palsies. Because the function of the cranial nerves is to carry information to the brain and to help the brain control sensory organs and muscles, the cranial nerves affect how we process information. When myelin is missing, nerve impulses are short-circuited. This impairment of cranial nerve function results in various kinds of thinking, perceiving, learning, and behavioral disorders. ... At the time of birth, myelination has just begun and in some nerves it doesn't begin until 8 months of age or later. Myelination proceeds at different rates in different neurologic areas for the next 15 years..."[17]

One author states that encephalitis is an almost expected reaction to vaccination, and there is much evidence to prove it.

"Encephalitis (brain damage due to swelling of brain tissue) is an almost given side effect to vaccine toxins according to most research studies on vaccine damages."[18]

Encephalitis also disturbs a child's myelination process.

"If myelin is prevented from being deposited or, once deposited, is removed (demyelination), the nervous system remains undeveloped and immature. The newborn infant, *especially if premature,* is clearly very much at risk. To be precise, a vaccination-associated encephalitis sometimes during the first year of life could easily interrupt the myelination process and thus cause neurological damage. ... The association between the post-encephalitic syndrome and demyelination or incomplete myelination of the brain seems quite secure, and the fact that encephalitis—*including that caused by vaccination*—can cause demyelination has been knowns since the 1920s!"[19]

"When the baby reacts to a DPT shot with a 'slight fever and fussiness' or 'drowsiness' for a few days, this may be, and often is, a case of encephalitis which is capable of causing quite severe neurologic consequences."[20]

The vaccinated versus unvaccinated homeschooled children studies also show that when a child is born prematurely *and* given all of the recommended vaccines, their chances of becoming developmentally disabled substantially increases. After seeing the exorbitant amounts of aluminum vaccines contain, and the implications of their encephalitogenic effects, is it any wonder that they have "14.5-fold increased odds"[21] of being developmentally disabled, compared to an unvaccinated child that was not preterm? Remember, without vaccination, this association between neuro-developmental disabilities and premature birth disappeared. We see over and over again, the younger the child, and the more exposure to vaccines, the more likely they are to have a negative reaction.

Here is another study that demonstrates the sagacity of vaccine refusal. This study sought to evaluate the amount of inflammation which occurs after vaccination by measuring C-reactive protein, which is described as:

"C-reactive protein [CRP] exhibits elevated expression during inflammatory conditions such as rheumatoid arthritis, some cardiovascular diseases, and infection. ... The highest concentrations of CRP are found in serum, with some bacterial infections increasing levels up to 1,000-fold. However, when the stimuli ends, CRP values decrease exponentially over 18–20 h, close to the half-life of CRP. CRP plasma levels increase from around 1 μg/mL to over 500 μg/mL within 24–72 h of severe tissue damage such as trauma and progressive cancer."[22]

The study tells us:

"Prospective observational study on 239 preterm infants at > or =2 months of age in the neonatal intensive care unit (NICU). Each infant received either a single vaccine or multiple vaccines on one day. CRP levels and cardiorespiratory manifestations were monitored for 3 days following immunization. ... Abnormal elevation of CRP level occurred in 85% of infants administered multiple vaccines and up to 70% of those given a single vaccine. Overall, 16% of infants had vaccine-associated cardiorespiratory events within 48 hours postimmunization.* In logistic regression analysis, abnormal CRP values were associated with multiple vaccines and severe intraventricular hemorrhage. Cardiorespiratory events were associated marginally** with receipt of multiple injections and significantly with gastroesophageal reflux. ... Precautionary monitoring following immunizations is warranted.***"[23]

*If these infants had not been under observation at a hospital but in their cribs, what would have been the result of these cardiorespiratory events?

**I wouldn't call 16% marginal, but I don't believe some children should be expected to suffer because of routine vaccinations.

***What about postponing immunization? Why is that never a suggested option?

After your child's doctor tries to persuade (or guilt) you into receiving the hepatitis B vaccine, they will do the same when your child is two months old, but with many more vaccinations. The rationale behind starting immunization at the age of two months was based on convenience rather than science.

"There is also a greater likelihood of completing the recommended series of injections if immunizations can be started in early infancy and coordinated with other reasons for the baby's being brought to the physician's office or clinic. For most practitioners, the age of 4 to 6 weeks would be a convenient time to start the series of immunizations against diphtheria, tetanus, pertussis and poliomyelitis. At this age the baby is starting his series of health supervision visits, and the mother is due for her traditional postpartum examination at about this time."[24]

Studies performed in Britain on the efficacy of vaccines in children aged 14 months were used to show that vaccines are somehow safe for two-month-old infants.

"The United States never conducted its own clinical trials to prove that the pertussis vaccine was safe and effective when given to children under one year of age. Instead, it relied...on data collected by Britain's Medical Research Council in clinical trials conducted in the 1950s for proof of the vaccine's safety and efficacy in newborns. But...these clinical trials, which included about 50,000 children, were not performed on two-month-old infants. The majority of the children in the trials were more than fourteen months old!"[25]

When your child is four months old, and then again at six, nine, and twelve months, the typical pediatrician will pressure you to let your child receive more of the same shots. Why do children have to receive the same injections repeatedly at such a young age? It seems as though perhaps their immature immune system might not even be capable of responding to the vaccines. In Chapter 4, when examining some of the medical literature on pertussis, a study was quoted which stated: "No protection by vaccination is demonstrable in infants."[26] One author wrote: "Did you know that 'all studies of efficacy of Hib vaccines admit that the vaccine is ineffective in children younger than 18 months'?"[27]

Your child's highest risk of dying from SIDS coincides with their receiving multiple vaccine injections into their small bodies. Why give these shots repeatedly when many of them have shown to be ineffectual, and their risks include permanent disability and death?

"Is it really worth the risk to expose two-month-old infants to a neurotoxin with all the above side effects in order to provide an imaginary temporary immunity, which wears off a few months later, requiring a booster shot? And all this for a disease that was disappearing anyway?"[28]

"Older data from the CDC used to list that approximately 10,000 babies die per year from what is called 'crib death,' or SIDS. 85% of these cases occur within the second to fourth month of life, and this just happens to coincide with the initial DPT and polio inoculations. <u>This is not a coincidence</u>."[29]

Many believe the polio vaccine is one of the most important for their children to receive. Polio has simply been renamed, not eradicated; and you will probably find it rather difficult to find an unvaccinated child diagnosed with any number of diseases that were once called polio.

"After the [polio] vaccine to this very day, spinal meningitis, a disease which is virtually identical to polio has been present in epidemic proportions."[30]

Whooping cough is another disease parents fear thanks to the misrepresentation of outbreaks broadcasted by the media. The truth is the medical literature has proven this vaccine's inefficacy, and its clear association with SIDS.

"Typically when pertussis vaccination is decreased in different parts of the world, as was done in 1981 by Sweden and the U.K., pertussis death rates actually dropped dramatically in both countries, England reporting the lowest death rates in recorded history. More whooping cough was diagnosed in these areas during this period however. ... In most cases throughout the world, doctors will refuse to diagnose a disease in which a person has been vaccinated for. This political bias alone is enough to alter statistics persistently showing a false yet favorable result for vaccines."[31]

"...when vaccination rates threaten to decline, physicians tend to diagnose pertussis 'every time a baby clears his throat,' to quote the late Robert Mendelsohn, M.D."[32]

Past "epidemics" have been examined, and reveal the truth of the breakouts hyped on the news.

"By autumn of 1982, forty-one cases had been reported, and the press release issued by the Maryland Department of Health was picked up by the local and national media. But when these cases were analyzed by J. Anthony Morris. Ph.D., an expert on bacterial and viral diseases, he found that in only five cases was there a definite laboratory diagnosis of whooping cough, and that all five of these children had been vaccinated. The 'epidemic' was much smaller than announced by the state health authorities and, such as it was, was occurring in a vaccinated population!"[33]

In our modern world, pertussis is a mild disease with few serious risks.

"Wolfgang Ehrengut, M.D., has reported...milder cases of whooping cough in West Germany, another country that stopped mass pertussis vaccination in the 1970s. 'Though vaccination had been performed on a low level, the fatalities due to pertussis had the lowest level ever recorded... In the last two to three years we had an increase in pertussis cases but (they were) of mild nature.' In a 1982 letter [Dr. Gordon] Stewart wrote, 'We have outbreaks about every four years. The last one in 1978/79 was more widespread, but the death rate was the lowest on record. In this city of over one million people (Glasgow), with less than 50 percent of children vaccinated, there have been no deaths since 1971, and we have no evidence of more severe or persistent complications due to pertussis. We do, on the other hand, have very clear evidence of some children having been hopelessly incapacitated following vaccination and of at least one unexplained death.' ... It appears that in the industrially advanced countries, the superior nutritional status of the populations and their better environmental circumstances are strengthening the host organism and thus taming the formerly lethal cause of whooping cough."[34]

In industrialized countries, more infectious disease diagnoses do not equate to more infectious disease deaths.

Many parents also fear measles, yet again because of the news media. There is much evidence that receiving the MMR vaccine at a young age can have terrible consequences for susceptible individuals, and according to some studies mentioned, school-aged children were still catching and spreading measles. Why should we give this vaccine to our young children when it will not protect them from measles just a few years later? According to the CDC: "The last measles death in the United States occurred in 2015,"[35] despite the "outbreaks" (as few as 3 cases) on the news. The source from which this information originated was last updated in April of 2019.

If one searches the VAERS database from January 2015 until March 2019 for "Death" and "Neonatal Death" in the United States from any measles vaccine (MM, MER, MEA, MMR, and MMRV), the results show that there have been 11 deaths. Because measles vaccines did not cause measles to disappear; it still spreads in highly vaccinated populations; there was only one death from measles between the years of 2015 to April of 2019, and in the same period there were 11 deaths attributed to measles vaccination; the government's own funded research identifies the gross underreporting of vaccine adverse events: **the greater risk of death lies with getting the vaccine rather than the disease**. When "…measles becomes a disease of immunized persons"[36] in highly vaccinated populations, what exactly is the benefit of this vaccine with its documented capability of killing?

As your child gets older, it does become safer for them to receive vaccines. Yet, there are even more reasons to decline, or at least substantially delay, vaccines for your children. In his book *Crooked: Man-Made Disease Explained*, Forrest Maready details the research which shows how injected metals can affect the human body in many ways. It started with the author noticing the many slightly asymmetrical smiles and eyes common among babies, and older individuals as well. His investigative research led him to discover a way that injected metals could be causing not only crooked faces, but also autism, and why it affects boys more than girls.

"Remember that I had been searching for a reason why the dorsal vagal complex was being immune activated. The dorsal vagal complex is that special location in the brain stem where many of the body's most important cranial nerves emanate from. Recently injected aluminum from vaccines moves around the body inside white blood cells via the blood and lymphatic system. Directly after injections, there is a higher amount of these aluminum-containing white blood cells because many of them have not yet found a resting place inside muscle tissue or been encapsulated inside granulomas.

I believe the reason acute cranial nerve damage is so common after vaccination is because the aluminum is free to travel to the dorsal vagal complex—as long as the body's immune system *signals* for it to go there. As I searched for answers as to why the immune system might signal for help at the dorsal vagal complex, I found two answers right away: pathogen invasion and tissue injury. Those two events—which happen anytime a child is vaccinated—would trigger immune activation within the dorsal vagal complex and cause a signal to go out to the body for help from white blood cells—many of which would unfortunately contain aluminum."[37]

"When I learned that these two specific events were mentioned as immune activation triggers, the answer was clear—this was why the dorsal vagal complex was getting so much damage. This was why cranial nerve problems were so common. The crooked faces that I had noticed seemingly everywhere were most likely because of the neurotoxin causing the cranial nerve lesions was introduced in such a way the body was practically begging for it to come wreck the part of your brain stem that controlled your face and your eyes. But other parts, too—your intestines, your mood, your behavior, and anxiety. Your ability to sleep well. Your ability to hear, taste, see and smell correctly. All of the illnesses associated with cranial nerves could be due to this one act—an injection of aluminum. … As it so happens, vaccines are administered in such a way they cause both triggers for immune activation in the dorsal vagal complex—and many of them simultaneously introduce the very ingredient you wouldn't ever want to get transported into that area. It seemed like the perfect crime—as if someone had purposefully designed this event to cause neurological damage. After having seen the rise in childhood neurological disorders over the past twenty years, I couldn't believe they weren't higher."[38]

It seems highly plausible if injected metals are drawn to a part of the body that the "…most important cranial nerves emanate from," this could cause nerve damage, which can manifest as a slightly crooked face or something much worse. If you want to see this phenomenon with your own eyes, look at pictures of children, and notice how many asymmetrical faces there are (or just google it). I was at the dentist's office looking at the "Cavity Free Club," when I saw for myself several noticeably "crooked" faces. Precious though these faces may be, I can't imagine parents wanting something in their child's body that is going to cause this nerve damage. Warning: once you start noticing crooked faces, they really are everywhere, and it's hard to not notice.

Aside from pathogens and injury, another trigger caused an immune response to this location: the "fight or flight" stress response that many children experience when they are forcibly restrained and given a painful shot, none of which they understand.

"A child will certainly cycle through the fight or flight stress response and the parents or nurse's restraint may send them into the next—something which will activate their dorsal vagal complex. This will happen to some extent whether the restraint is required or not, simply due to the other two triggers happening—pathogen invasion and tissue injury. If you add the third one due to immobilization and restraint, you will end up with a massively activated dorsal vagal complex. ... No one enjoys this process—doctors, nurses, parents or children, but it was always assumed that 'the end justifies the means.' The mechanisms described in the polyvagal theory, and their potential impact on the safety of injecting metals into children's bodies may force us to reconsider that statement."[39]

While this is a theory, it is certainly worth scientific investigation since modern medicine "just doesn't know" what causes autism. If this one man could piece all of this together, imagine what would happen if the entire medical community would objectively search for the truth; even if it meant their cash cow "routine vaccinations" wouldn't be nearly as profitable.

A high aluminum content has also been noted in the brains of autistic children.

"Autism spectrum disorder is a neurodevelopmental disorder of unknown aetiology. It is suggested to involve both genetic susceptibility and environmental factors including in the latter environmental toxins. Human exposure to the environmental toxin aluminium has been linked, if tentatively, to autism spectrum disorder. Herein we have used transversely heated graphite furnace atomic absorption spectrometry to measure, for the first time, the aluminium content of brain tissue from donors with a diagnosis of autism. ... The aluminium content of brain tissue in autism was consistently high. ... These are some of the highest values for aluminium in human brain tissue yet recorded..."[40]

One of the study's authors postulates:

"'One has to wonder why aluminum in the occipital lobe of a 15-year-old boy with autism would be a value that is at least 10 times higher than what might be considered acceptable for an elderly adult?' said Christopher Exley PhD, Professor in Bioinorganic Chemistry and author of the study. Another ground-breaking study by Exley and his team, published earlier in the year, identified similarly high levels of aluminum in the brains of individuals who died of familial Alzheimer's disease. While the aluminum content of each of the five brains was extraordinarily high, it was the location of the aluminum in the brain tissue that served as the standout observation. ... According to the researchers, there was clear evidence of inflammatory cells heavily loaded with aluminum entering the brain via the meningeal membranes and the blood-brain-barrier. ... 'The research does not infer that Al is a cause of autism,' said Exley, who also authored 'Human Exposure to Aluminum.' 'But these very high concentrations of a neurotoxin in brain tissue are not going to be benign and will contribute to neurodegeneration in affected tissues.'"[41]

Maready also theorized that because men experience the "fight or flight" response much more so than women, and boys typically receive another tissue injury before they leave the hospital —immediately after an injection of the aluminum-containing hepatitis B vaccine—they are much more vulnerable to this nerve damage that can manifest in many ways.

"If your children require restraint in order to administer shots, then in my opinion, they are too young to be safely immunized. This restraint will trigger their dorsal vagal complex in such a way as to signal the metals that were just injected into their bodies straight to vulnerable parts of their brain. Wait until they are old enough to understand, then make a decision with them regarding vaccines."[42]

Another vaccine many parents skip is the HPV vaccine, which does not use an aborted fetal cell line but is associated with many adverse reactions. One of the best resources for parents on vaccines is a book called *Miller's Review of Critical Vaccine Studies* by Neil Z. Miller. In this book, 400 studies are summarized for parents. A section on the HPV vaccine shows many studies link this vaccine with lupus, autoimmune disease, nervous system damage, and reproductive problems. Several of the safety studies mentioned earlier which utilize a false placebo are testing the HPV vaccine. When an actual placebo is used, the results are not favorable to this vaccine. One study "...sought to evaluate the effects of Al adjuvant and the HPV vaccine Gardasil versus the true placebo on behavioral and inflammatory parameters in female mice. ... It appears that Gardasil via its Al adjuvant and HPV antigens has the ability to trigger neuroinflammation and autoimmune reactions, further leading to behavioral changes."[43]

One last item to discuss is tetanus:

"Tetanus is very rare in the United States and affects more adults than young children. It may be a danger if one steps barefoot on a rusty nail in a horse stable and has a weak resistance to disease. This fails to justify the insistence of most emergency medical departments to administer tetanus shots to everyone without any wound at all. The best prevention for this disease is proper wound hygiene, including allowing the wound to bleed (for cleansing) and to be in contact with the air (the tetanus microorganism is anaerobic; that is, it develops in the absence of air). Tetanus is treatable with antibiotics."[44]

Tetanus has also been cured by—you guessed it—vitamin C:

"Jahan et al. (1984) conducted a simple study on the effects of vitamin C in the treatment of tetanus. In 31 patients aged from one to 12 years, a daily intravenous vitamin C administration of 1,000 mg was given in addition to antitoxin [antibody treatment]. Although none of the vitamin C-treated patients died, almost 75% of those who did not receive the vitamin C did."[45]

Imagine if one of your children was in this study, and vitamin C greatly determined whether they lived or died. Why does modern medicine not use this safe, nontoxic method?

"Vaccine Hesitancy" is considered a top-ten global health threat. Personally, I believe it's a threat to the pharmaceutical industry, not the people. The historical fact is that vaccines did not cause infectious diseases to disappear. When reading through smallpox literature, a consistent theme exists: the unvaccinated were better protected, and less likely to suffer and die because of smallpox. This still holds true today with other infectious diseases. Children in many other countries receive substantially less vaccines, and: "Their citizens are not dying from infectious diseases."[46]

"Well, now we all know what happened when smallpox vaccination was discontinued: exactly nothing. No epidemics of smallpox followed, not even due to potential importations via a vastly intensified movement of the travelling public. Equally, there will be no epidemics of these diseases when routine diphtheria and tetanus vaccination of infants is discontinued. When whooping cough vaccination is stopped, as it was in Sweden, Italy and the former West Germany, the incidence of pertussis will diminish or will continue at the same rate as present, but with some important differences. The age distribution will return to normal. Only about 10% of whooping cough cases will occur in babies below the age of one and the rest in those children several years old who can take the disease well. The tragic toll of vaccine-induced diseases will stop abruptly and the bizarre modern entity of cot death, as we know it now, will disappear (as it vanished from Japan after 1975)."[47]

"Again, medical literature brings evidence upon evidence that vaccinated children contract the very diseases against which they are vaccinated at the same as or at a higher rate than unvaccinated children."[48]

"The modern plague —"vaccination"— is destroying immune systems in the industrially developed countries. Potent poisons and foreign antigens are being daily introduced by injection…into the bloodstream of tiny babies. These are followed by potent immunosuppressive chemicals called, quite ironically, 'medications.' … History is certainly repeating itself; humanity has not really become wiser or more scientific for all its technical progress."[49]

Just as smallpox vaccination did not stop smallpox, modern vaccination will never eradicate, nor ameliorate, any disease. This is an unpopular fact, but popularity has no bearing on truth.

"The demonstration appears conclusive, that smallpox will never be eradicated, nor in any material degree mitigated in its severity, by vaccination or other contamination with blood poison, I expect no favor or even lenient treatment, for uttering this conviction. The 'dupe of a deep delusion,' seldom cherishes kind feeling toward the individual who disturbed his quiet. He prefers to be left alone in his mistake. It is easier to break the thermometer than to acknowledge the true state of the atmosphere; to stone the prophet than to heed his words. … Mankind, even though they may not choose to be deceived and led by the nose, are much annoyed when they are required to think and act on their own responsibility. The charlatan has a larger following than the philosopher, and is certainly a great deal more noisy."[50]

Please wait and pray; and do not receive all of your information about vaccines from those who profit from them.

"Perhaps it would be prudent not to get all the information on that subject from those making a living selling vaccines. … Nothing in the universe is more delicate than the infant's brain and immune system, as they struggle into existence. Vaccines have the undisputed ability to damage both. … What is the common lament of those horrified parents? 'I wish I'd known.' That's what they all say, once they find out, once it's too late. I wish I'd known… This is your chance to know."[51]

Even if your child receives every single vaccine, they *might* be protected from a very, very small number of possible illnesses. Our God-given immune systems, when given proper nutrients and not stifled by unnatural interference, can handle almost all of these germs.

"We have vaccines for just a tiny percentage of known viruses and bacteria—probably less than .01%, yet you could probably not name a single one. You are afraid of the one listed on the childhood vaccine schedule, but not the others. …vaccines can interfere with proper immune system function and [we] know that when optimum health and nutrition are present, the immune system can take care of nearly any infection—even the ones for which vaccines don't exist."[52]

Man is incapable of fully understanding the immune system, and our government refuses to properly investigate the negative short- and long-term effects of vaccination: "...it's no metaphor to say our children are the lab rats."[53]

"That complex biological symphony of interrelated allergic responses, antibodies, antigens, self recognition, cell response, etc — about which we still have only the vaguest knowledge — struggles its way into existence during the early years of the child's life. It needs to help, no interference, no enormous experimental toxic load, especially one so politically contrived, in its fight to survive."[54]

"Perhaps the darkest consequences of all the foregoing is that most of us have lost confidence in the inner curative power of Nature - the body's inborn wisdom. A hundred media snippets a day, week after week, year after year, have undermined our ability to even consider the notion that 99.9% of infants may be perfect as they are. Or that their pure blood is the most sacred medium in the universe, the crucible in which the human genome itself was meant to be safeguarded and passed on from age to age. Or that the immune system can only develop to its full potential if left to its own devices, largely unknown to human science."[55]

I don't believe in the inner curative power of nature, but rather that the loving, eternal God of the Bible designed our bodies to heal without man's intervention. Of course there are exceptions—we do live in a fallen world. Why is it those who believe in a benevolent Creator doubt the wisdom He manifested in nature more so than those who believe everything is random and popped into existence by chance? Why is it that this author believes an infant's blood is "...the most sacred medium in the universe..." when we Christians have a Bible that declares the life of flesh *is in the blood*? What are the consequences of putting disease into what God calls the life of our flesh? What are the consequences of injecting aborted fetal DNA into the LIFE of our children? There is a terrible disease that cannibalism begets called kuru; do we believe injecting human DNA will be free from any ramifications? The life of our children's flesh is a "...medium not intended..." for foreign human DNA, just as it is not the proper medium for any single ingredient found in a vaccine.

"The most serious, even fatal, disorders may be provoked by the injection of living organisms into the blood; organisms which, existing in the organs proper to them, fulfil necessary and beneficial functions—chemical and physiological—but injected into the blood, into a medium not intended for them, provoke redoubtable manifestations of the gravest morbid phenomena..."[56]

Just a few short years ago, researching this subject would have been much more difficult. Today, there is abundant information available to inform yourself. These innocent, newborn children are completely helpless; parents are their only protectors. Parents must do what God intended and educate ourselves ("*Prove all things*"), so that we can properly protect our children.

It has been observed that doctors quote disease statistics from third world countries, where filth and malnutrition make infections much more common and deadly, to try to convince parents to vaccinate. One mother of a child who had already experienced two negative reactions to the DPT vaccine postponed giving her son the third dose. She stated that she "dreaded giving it to him. I didn't want him to have another shot, but the doctors really pressure you. I've always been very respectful of doctors and medicine. You just want to do the right thing."[57] Her son received another shot, and had another reaction. He went on to have constant ear and respiratory infections, and his mother believes he has a learning disability. Physicians are *trained* to talk you into vaccinating your child; parents must not submit to their fear-mongering tactics.

"Fear and intimidation are powerful weapons. These are the weapons of mental torture applied to make parents submit their babies to this barbaric, unscientific and totally unnecessary procedure. However, it is the parents who have all the power over vaccination in their hands. ... Above all, parents have a moral and legal obligation to ensure health and good prospect of a full life for their children. Their loyalty should at all times be with their children and not with their peers or the medical system. ... The medical system in the developed countries has become totalitarian and highly politicised. It lacks the philosophy of healing and true knowledge of the physiology of the human body. It is little more than a big business which is mandating their procedures and enforcing their medications."[58]

For the sake of our children, we must arm ourselves with information. Parents, it is imperative to find a doctor who respects your right to choose what is best for your child. Be warned, some parents have been reported for "neglecting" their children because they have not been vaccinated.

If you still want to vaccinate your children, please first find "incontrovertible proof" that these vaccines will actually protect—"This is your child we're talking about here."[59] It is unwise to be so trusting of a world that God called evil (Gal 1:4), or to trust products and policies made by fraudulent and deceitful men who fully support the taking of human life before birth, and have proven to be motivated by their love of money. Their methods to manipulate: "...cannot deceive [anyone] who is earnestly endeavoring to ascertain the truth concerning this important question."[60]

Proverbs 12:15 **The way of a fool is right in his own eyes: but he that hearkeneth unto counsel is wise.**

Proverbs 11:14 **Where no counsel is, the people fall: but in the multitude of counsellors there is safety.**

Please research and pray for wisdom while you are deciding how you're going to keep your children healthy. Many voices give their opinion on this issue, and I'm willing to at least hear them. At the end of the day, any advice that goes against God's Word is of no value. In the book of Job, his friends were trying to help him by telling him what *they* thought. Any person's opinion is only as good as their information. It was not until God spoke that Job was given the truth. God's words must be the standard.

Job 13:2 **What ye know, the same do I know also: I am not inferior unto you.** 3 **Surely I would speak to the Almighty, and I desire to reason with God.** 4 **But ye are forgers of lies, ye are all physicians of no value.**

"Infectious disease specialists use the term 'refuser' pejoratively, to cast aspersions on those who are intelligent, informed, and strong willed enough to 'refuse' the corporate and government assault on our families' bodies with their ineffective, unsafe, and toxic vaccines. ... After losing a daughter [to a hepatitis B vaccine reaction] and investigating the so-called science behind our current vaccines, *I am a refuser*. Medical authorities and uninformed parents want to turn that informed decision around and paint people like me as a menace to society. The VAERS reports show that *adverse reactions to vaccines are the real menace to society*."[61]

"We need not employ a Satan to cast out Satan, but only the 'finger of God.'"[62]

"Our children face the possibility of death or serious long term adverse effects from mandated vaccination programs that aren't necessary, or that have very limited benefits."[63]

—Jane Orient, MD Executive Director of the Association of American Physicians and Surgeons

"It has never ceased to amaze me that parents allow their children, the most precious aspects of their lives, to be injected with substances they know nothing about."[64]

"The fact that vaccines create cancer, autism and other degenerative neurological disorders is so blatantly obvious from the perspectives of both science and law, that the issue is no longer up for debate."[65]

"If you do vaccinate, you had better be comfortable about the need for each vaccine, because every time you vaccinate your child, there is a risk of severe injury and death."[66]

Psalm 27:14 **Wait on the LORD: be of good courage, and he shall strengthen thine heart: wait, I say, on the LORD.**

Many speak things that are "*...of no value,*" and with so many different professional opinions, how do we know what's true?

"When a man's knowledge is not in order, the more of it he has, the greater will be his confusion."—Herbert Spencer

In the Bible, Daniel was told in the last days knowledge would increase (Dan 12:4). Man has much knowledge, and also much confusion, because man has no standard. Christians, however, have the source of ultimate truth. We will turn to the Scriptures to see what can be gleaned about health. Man "*...darkeneth counsel by words without knowledge*" (Job 38:2), but God is a God of truth (Deu 32:4), and His "*...counsels of old are faithfulness and truth*" (Isa 25:1).

Isaiah 50:7 **For the Lord GOD will help me; therefore shall I not be confounded: therefore have I set my face like a flint, and I know that I shall not be ashamed.**

Proverbs 24:13 **My son, eat thou honey, because it is good; and the honeycomb, which is sweet to thy taste:** 14 **So shall the knowledge of wisdom be unto thy soul: when thou hast found it, then there shall be a reward, and thy expectation shall not be cut off.**

The Bible says many good things about waiting on the Lord; in contrast, let's see what happened to Israel when they chose not to wait on God:

Psalm 106:13 **They soon forgat his works; they waited not for his counsel:** ... 24 ...**they believed not his word:** ... 28 **They joined themselves also unto Baalpeor, and ate the sacrifices of the dead.** 29 **Thus they provoked him to anger with their inventions: and the plague brake in upon them.** ... 35 [They] **were mingled among the heathen, and learned their works.** 36 **And they served their idols: which were a snare unto them.**

Chapter 7

Seek to the LORD

Proverbs 4:20 **My son, attend to my words; incline thine ear unto my sayings.** 21 **Let them not depart from thine eyes; keep them in the midst of thine heart.** 22 **For they are life unto those that find them, and health to <u>all</u> their flesh.**

Proverbs 3:6 **In <u>all</u> thy ways acknowledge him, and he shall direct thy paths.**

Hopefully, you now have enough information to make an informed decision. We might be inclined to believe that an unvaccinated child would require extra care to keep them well; in reality this is a misconception, and the opposite is true. It is the vaccinated children that would require the extra care—in dealing with illnesses or infections that these children experience much more frequently, or to help their bodies detox from vaccine exposure. Even with this knowledge, the decision to forgo or delay vaccines is unnerving at first because we believe our children will be "unprotected." It is those who have had their immune systems changed, and therefore weakened, through vaccination that are less protected. I want to show you what the Bible truly associates with health to help give you some peace of mind.

We live in a much cleaner environment than man ever has, and because we are not malnourished (from a *lack* of food), we do not have to worry about many infectious diseases being deadly in our modern world. But it is very easy to not eat properly and promote sickness as a result. Diet is truly one of the most important aspects of human health (see Psa 103:5; Pro 23:1-2). This section will focus on basic principles we can learn from the Scriptures that can greatly benefit our health and look at some of the shortcomings of what man calls medicine.

Ecclesiastes 7:17 **Be not over much wicked, neither be thou foolish: why shouldest thou die before thy time?**

The members of Christ's body (1Co 12:27), the New Testament Church (Col 1:18), are not promised health, wealth, or problem-free lives. However, if foolishness can bring about our death "*before*" our time *as the Bible says,* we must go to God's Word for wisdom. Diet is a subject in which the Church has freedom, where every man should be "*fully persuaded in his own mind*" (Rom 14:5), and all things are lawful—*but* not all things are expedient (1Co 10:23). Let me briefly tell you what I believe is most expedient in regard to our health, according to the rightly-divided Word of God. I do not expect everyone to agree with me on this issue, and according to the Bible, we are clearly not supposed to judge Christ's Church over meat and drink (Col 2:16); I simply want to present this in a way you may never have heard before.

Proverbs 30:8 **Remove far from me vanity and lies: give me neither poverty nor riches; feed me with food convenient for me**:

The word translated to convenient comes from the Hebrew word which means "something prescribed or owed, a statute." Eating foods as they are in nature (as God made them), would undoubtedly be the prescribed, convenient foods for us to eat.

Psalm 103:5 **Who satisfieth thy mouth with good things; so that thy youth is renewed like the eagle's.**

As has already been stated, man cannot do anything better than God; changing food from its natural state to a processed product is to change its God-given design. We truly have so much influence over whether we are healthy by eating properly—eating foods as God made them.

"…if the body were maintained in an alkaline condition, this would 'immunize' a person against infectious diseases."[1]

"'Even the World Health Organization has conceded that the best vaccine against common infectious diseases is an adequate diet. Despite this, they made it perfectly clear to us that they still intended to promote mass immunization campaigns,' Drs. Archie Kalokerinos and Glen Dettman tell us. 'Do we take this as an admission that we cannot or do not wish to provide an adequate diet? More likely it would seem, there is no profit in the constituents of an adequate diet for the pharmaceutical companies.'"[2]

"…infectious disease and malnutrition are so inextricably interwoven…"[3]

"It is almost unbelievable how mild these childhood diseases can be when no malnutrition exists."[4]

"…a diet consisting of organically grown fruits and vegetables and little animal protein [plays] an important role in preventing disease."[5]

"…if the body were kept in an alkaline condition by the addition of carrots, celery, lettuce and the like, each day, there would be no need for vaccinations."[6]

Many western minds scoff internally at the thought of nutrition preventing disease; please remember the account of Dr. Akizuki, whose diet protected him from radiation sickness after the *atom bomb* that exploded over Nagasaki.

1 Corinthians 1:27 **But God hath chosen the foolish things of the world to confound the wise; and God hath chosen the weak things of the world to confound the things which are mighty;** 28 **And base things of the world, and things which are despised, hath God chosen, yea, and things which are not, to bring to nought things that are:** 29 **That no flesh should glory in his presence.**

While we, thankfully, do not have to worry about malnutrition the same way many do in other parts of the world, we are still suffering from many chronic diseases because of our low-nutrient and highly-processed diets.

"The *overcrowding, lack of sanitation, and malnutrition associated with poverty* are well-documented immune suppressants. Less well known, perhaps, is the malnutrition of the affluent caused by the high intake of refined, chemicalized, overcooked food. These processes not only damage the food so that it doesn't deliver the nutrients needed by the body but leave toxic metabolic residues in the body. Did you know that sugar consumption depresses immune function with remarkable rapidity? As little as 100 grams of sugar reduces immune function by 50 percent within the hour."[7]

We will all die eventually, no matter how well we eat. I do not want my death to be **before my time** because of my own *refusal of instruction* (Pro 15:5)—foolishness—as the Bible says. Also, with the time I do have, I want to be healthy to accomplish something for the Lord.

I understand that we are not supposed to command any food to be avoided, and all things are good and nothing to be refused (1Ti 4:3-5). Yet, to take these verses from Timothy and apply them to the gluttonous ways of our culture is to not take the Bible as a whole. The Bible warns us about gluttony's dire consequences in Proverbs 23:2: "*And put a knife to thy throat, if thou be a man given to appetite.*" The last verse clearly tells us there is a certain way to go to an early grave—to be given to appetite. Ezekiel 16:49 details about some of Sodom's lesser-known sins: "*Behold, this was the iniquity of thy sister Sodom, pride, fulness of bread, and abundance of idleness was in her and in her daughters, neither did she strengthen the hand of the poor and needy.*" The Bible also says not to be deceived—a man will reap what he has sown (Gal 6:7). If what we sow is based on man's flawed ideals or our own ignorance, which are both in opposition to God's word, is it so surprising we reap disease and premature death?

My simple advice to you would be this: Eat food the way God made it as much as possible. One of the main reasons we want to take care of our bodies is to keep them strong to serve God. In our modern world, we are constantly exposed to things that can harm our bodies, which is why eating well is very important to our long-term health. A proper diet helped protect from radiation sickness in Nagasaki; our bodies have an amazing proclivity toward health and healing, but proper nutrition is required. Eating healthy does not have to be an all-or-nothing scenario. Simply put: eat well most of the time, and you will be much healthier. The Bible does not promise health or wealth to New Testament Christians, but it does contain principles that will benefit us in these areas when we apply them. We should go to God's Word and learn what we can on any matter. I understand that the

Church is not Israel, but the Scriptures tell us that the things which happened to Israel were for *our* admonition.

Psalm 106:14 **But lusted exceedingly in the wilderness, and tempted God in the desert.**

Psalm 78:18 **And they tempted God in their heart by asking meat for their lust.**

Psalm 106:15 **And he gave them their request; but sent leanness into their soul.**

1 Corinthians 10:6 **Now these things were our examples, to the intent we should not lust after evil things, as they also lusted. ... 11 Now all these things happened unto them for ensamples: and <u>they are written for our admonition</u>, upon whom the ends of the world are come.**

In the course of a day, do you think more about food that will fill your belly, or about God and His Word that will fill your soul? Christians must take care to not be like many of the Israelites, who confessed love with their mouths, but their hearts were after covetousness (Eze 33:31-32). I am a great offender of this, and am really not as spiritual as I might sound by asking this question. *But I want to be better*; I do not want to be so driven by my desire for food and other earthly goods that I'm not serving God to my full potential. I want to have money to give to those in need instead of spending every penny on things for myself that I don't need, that only provide momentary enjoyment, and are void of even an infinitesimal speck of significance in the fathomless eyes of the eternal God to whom I will have to answer (Rom 14:12). We should not be so focused on our lust for things; they're all going to burn (2Pe 3:10). These thoughts might seem like they don't belong in this section about health, but our lifestyle, as well as our "thinkstyle," affects our health. The common preoccupation with the desire for *more*, instead of the Biblical attitudes of contentedness (1Ti 6:6) and thankfulness (Col 2:7), does not promote health.

Ecclesiastes 10:17 **Blessed art thou, O land, when thy king is the son of nobles, and thy princes <u>eat in due season, for strength, and not for drunkenness</u>!**

I'm not saying it is wrong to enjoy food, quite the opposite: please enjoy it and *be thankful*. I'm also not suggesting you live off of fruit, seeds, and herbs like Adam and Eve. I'm suggesting to examine yourself and be sure you are not consumed by lust for food, or whatever else, like the Israelites were. God let them have what they wanted but sent "*...leanness* <u>*into their soul*</u>." Lusting after things will not help one's spiritual health, which will affect physical health as well. It is not wise for us to be constantly focused on food—to "*eat...for drunkenness*." Trusting God's Word and eating "*for strength*" to do what God has told us to do, is much healthier than constantly looking for your next fix—food, earthly goods, or experiences—the lust of the flesh, the lust of the eyes, and the pride of life (1Jn 2:16; Gen 3:6).

I also want to make a few quick comments about fasting and exercise. Jesus expected His believers to fast after His ascension (Luk 5:34-35). Anna *served* God with fastings (Luk 2:37), and Christ said a combination of fasting and prayer could accomplish things that otherwise could not be done (Mar 9:29). We are not commanded to fast, and *some people cannot or should not fast*. If we skip a meal occasionally and instead pray to God and help others instead of always helping ourselves, it will benefit us physically and spiritually. The Bible says our soul is humbled through fasting (Psa 35:13), and while God resists the proud, He gives grace to the humble (1Pe 5:5). If fasting with the right heart attitude can do all these things, why would we not utilize such a simple method? To be clear, I'm talking about *short* fasts, as in a day or less. (If Anna served with fastings night and day, they must have been short.)

Isaiah 58:6 **Is not this the fast that I have chosen? to loose the bands of wickedness, to undo the heavy burdens, and to let the oppressed go free, and that ye break every yoke?** 7 **Is it not to deal thy bread to the hungry, and that thou bring the poor that are cast out to thy house? when thou seest the naked, that thou cover him; and that thou hide not thyself from thine own flesh?** 8 **Then shall thy light break forth as the morning, and thine <u>health shall spring forth</u> speedily: and thy righteousness shall go before thee; the glory of the LORD shall be thy <u>re</u>reward.**

As for exercising, we all know it is physically beneficial (notice the words *"for thy sake"* in Gen 3:17). Exercise can help one's mental health as well. In the Bible, the only time the words "bodily exercise" are used, Paul writes that it profits little, but godliness is profitable unto all things (1Ti 4:8). I believe it's important to note that bodily exercise would be defined differently in Paul's time than ours—he would have walked nearly everywhere he went (See Act 20:13), which is much different than how we live today. The Bible has much more to say about not being slothful than it does about bodily exercise profiting little, and it says much more about godliness than it does about either slothfulness or exercise.

It is much more important to focus on spiritual things than physical; what comes out of our mouth is more important than what goes in (Mat 15:16-20), but few realize the effects food can have on us. Remember, God sent leanness *into* their souls when He gave the Israelites what they lusted after. I'm only presenting this information to supplement the knowledge many already have about the main focus of the Christian life—to be focused on spiritual things. He that is spiritual judgeth all things (1Co 2:15), *including* whether or not what they eat is for their own lust, or for the glory of God (1Co 10:31).

I'm not suggesting you never eat food for enjoyment (see Deu 12:7) or that you should never eat processed or fast food. I'm suggesting that you don't eat it every day of your life, and instead eat foods the way God made them as much as possible, and eat them with thankfulness. Be aware that many processed foods are hard to avoid because you might be literally addicted to them. While the Apostle Paul said that all things were lawful, he also said he would not be brought under the *power of any* (1Co 6:12).

"Calories are condensed in the same way plants are turned into addictive drugs like opioids and cocaine: concentration, crystallization, distillation, and extraction. They even appear to activate the same reward pathways in the brain. Put people with 'food addiction' in an MRI scanner and show them a picture of a chocolate milkshake, and the areas that light up in their brains are the same as when cocaine addicts are shown a video of smoking crack or when alcoholics are given a whiff of whiskey."[8]

Limiting processed foods will have nothing but positive effects on your health, and more importantly, you will avoid processed foods that utilize aborted fetal cell lines during their research and development (not in the final products themselves). *Children of God for* Life has a complete list of all aborted fetal products on their website.[9] Yes, somehow in this messed-up world, there are aborted fetal cell lines used to develop flavor profiles (as I understand), and even the subject of food brings us back to aborted fetal cell lines. (The wicked *"search out iniquities; they accomplish a diligent search,"* Psa 64:6. Man is as wicked as he can be, Jer 3:5.) I'm not sure how common this practice is, and I'm not even sure this is still happening—*Children of God for Life* has done much research, and it is all available on their website. Here is a quote from a study published in 2010 called "Molecular mechanism of the sweet taste enhancers." The history of the cell line mentioned in this study from their website (cogforlife.org) tells us "HEK (human embryonic kidney) 293 – the number of aborted fetal experiments prior to establishing the cell line."[10]

"Positive allosteric modulators of the human sweet taste receptor have been developed as a new way of reducing dietary sugar intake.* … As previously described, hT1R2-1/hT1R3 stable cell lines were generated by transfecting linearized pCDNA3.1/Neo-derived T1R2-T1R1 and pCDNA3.1/Zeo-derived (Invitrogen) T1R3 vectors into a $G_{16gust25}$ cell line,** an HEK 293 line stably expressing the chimeric $G_{16gust25}$ protein. … Resistant colonies were expanded, and their responses to sweet and umami taste stimuli were evaluated by calcium imaging.[11]

*It's pretty pathetic the lengths that man will go to trick himself into eating healthier. I don't care how good something tastes, if this was the method used to create a certain flavor, *"while the world standeth,"* (1Co 8:13) I'm not eating it. See also 1Ch 11:18-19; Psa 141:4. (Realistically, we have all probably ingested products developed this way in ignorance.)

**Imagine what could be accomplished if the people who understand this sentence didn't waste their lives with aborted fetal research (Gen 11:6).

We also learn from the Bible our thinking influences our health. God only calls two things medicine: a merry heart (Pro 17:22), and a leaf (Eze 47:12). These two medicines could be used in many different ways by many different methods to promote healing, but we will keep it simple. The importance of a Biblical attitude is **crucial to**

our health, as God said. We have a great amount of influence over our health and quality of life by *thinking* properly.

Philippians 4:8 **Finally, brethren, whatsoever things are true, whatsoever things are honest, whatsoever things are just, whatsoever things are pure, whatsoever things are lovely, whatsoever things are of good report; if there be any virtue, and if there be any praise, think on these things.**

"The consequences of stress...can trigger the immune system in deleterious ways that no change in food or exercise regimen could touch. ...stress can exist for years—even decades—wearing down the immune system to nearly nothing."[12]

"There is 'overwhelming evidence that the mind is a key controlling factor...in virtually all disease,' [says] Edgar Mitchell, former astronaut and founder of the Institute of Noetic Sciences..."[13]

"...lifestyle and 'thinkstyle' are primary determinants of health or its lack."[14]

Proverbs 17:22 **A merry heart doeth good like a medicine: but a broken spirit drieth the bones.**

"The power of the mind and emotions to create and correct disease states has been widely discussed. 'Every thought you have creates changes in the body,' according to Dr. Norman Shealy. He says that joy increases the strength of the immune system, and sadness or depression decreases the strength of the immune system."[15]

Proverbs 12:18 **There is that speaketh like the piercings of a sword: but the tongue of the wise is health.**

Proverbs 16:24 **Pleasant words are as an honeycomb, sweet to the soul, and health to the bones.**

"Robert Ingersoll once remarked that had he been God he would have made health contagious instead of disease. 'When shall we come to recognize that health *is* as contagious as disease,' an Indian contemporary responded. Virtue is as contagious as vice, and cheerfulness is as contagious as moroseness, he continued. When the mental images of a community of minds are filled with pictures of health rather than disease, the possibilities for the realization of health become practically limitless. ... 'Each patient has the potential of being his own healer."[16]

For all that we spend on health care, are we any healthier? Are modern medicine's methods in treating chronic disease effective?

"The more I read and study, the more I believe that medicine is heading in the wrong direction. I am not denying that modern medicine has done some marvelous things - acute emergencies and trauma management are unparallelled, but that is where its usefulness ends. Modern medicine is trying to control nature. This is totally wrong - we should be working with nature."[17]

If the word nature or natural bothers you, think of it this way instead: the way God made it, or the way God designed it. If we do not work with God's designs, but against them, it will not benefit our health. When man ignores God's Word and monopolizes "medicine," it will not be what it should be. I am not against modern medicine: where it agrees with Scripture it's certainly effective, but where it does not align with the Bible, it is simply wrong, as it has been many times in the past.

"As a retired physician, I can honestly say that unless you are in a serious accident, your best chance of living to a ripe old age is to avoid doctors and hospitals and learn nutrition, herbal medicine and other forms of natural medicine unless you are fortunate enough to have a naturopathic physician available. Almost all drugs are toxic and are designed only to treat symptoms and not to cure anyone."
—Alan Greenberg, M.D.

"When a group in power has a vested interest in sickness, is it too much to suggest that their approaches to the solution of this problem will be as expensive, circuitous, and complex as possible?"[18]

"If you choose to go to a hospital, go into hospitals with your eyes wide open. Medical care, like so many other things, is problematic; it's not a sure thing. Patients must understand hospitals are hazardous and medical care is a dangerous enterprise - you must be willing to put a considerable amount of energy into self-protection."[19]
—Lowell Levin. M.D., Professor at Yale School of Medicine

God designed *simple* things to keep man healthy. God says He is the only one who heals, and He has designed our bodies to heal. If your body has what it needs, it will heal itself, unless we have damaged our bodies beyond their capacity to heal. The simple things that promote healing are healthy food, clean water, sunlight, fresh air, exercise, etc. This world wants to make medicine highly profitable. These two premises are completely opposed to one another. We have a medical system that puts profit before life, and a benevolent God who freely gives life and all things good. Medication and surgery should certainly be used when they are necessary; but taking a toxic drug for every single bodily complaint does not lead to good health.

While it is true there are many people who rely on medications to live, it's important to realize, based on the evidence presented in medical literature, their childhood exposure to vaccines could very well be the reason they have these chronic health conditions in the first place. Malnutrition in the womb can also cause physical problems that cannot be remedied.[20] Nutritional deficiencies spanning years also contribute to many of the diseases we would consider to be largely caused by genetics. Proper nutrition and limiting our exposure to toxins are truly proving to be more important to human health than previously understood.

God said in Genesis 11:6 that the people of Babel were one, *"...and they have all one language; and this they begin to do: and now nothing will be restrained from them, which they have imagined to do."* When it comes to certain things, mankind is of one mind and one language. I believe the love of money is one of those things. If man would put aside the goal of monetary gain and work together to find disease cures and answers, I believe we would know much more than we do. The reality is the main goal of this research is profit. The medical community truly does want to cure these chronic diseases that plague our society, but *only* if these cures can be *patented and sold*. But this is not how God designed our world. Health is not discovered by experimenting on humans and animals in a lab—although we can learn through morally-obtained materials.

Man will never discover the truth that he is capable of knowing when he rejects God (Rom 1:28-29). People have been cured of many cancers and other "incurable" diseases by using natural, nontoxic methods, and these are much cheaper than what modern medicine offers. When a doctor cures someone of a disease using methods not sanctioned by Big Pharma, *they are actively suppressed and/or persecuted*. The pharmaceutical industry wants to own health and healing, but this is impossible. Man cannot own what only God can do.

For hundreds of years, man has profited from the sale of metals and disease as medicine, neither of which align with what God calls medicine. Mankind's rejection of God's laws while attempting to find cures with profit as their sole motivating factor has led to many chronic diseases. Man is looking for the answers in the wrong way by working against God's designs. *Man is searching for truth that opposes*

God's Word; man is searching for truth where it does not exist. This is a quixotic notion, to say the least. While this world's medical systems maintain their efforts of tilting at windmills, I'm just going to trust God's Word.

While Jesus did say they that are whole do not need a physician—He also clearly states those who are not whole **do** need a physician (Luk 5:31). Physicians certainly can aid the healing process (Luk 4:23a; Jer 8:22). This does not mean that a physician who ignores God's Word and suggests things such as aborted fetal cell lines to bring health is the kind of physician to whom Jesus is referring.

2 Chronicles 16:12 **And Asa in the thirty and ninth year of his reign was diseased in his feet, until his disease was exceeding great: yet in his disease he sought not to the LORD, but to the physicians.**

This verse is not used to imply it is wrong to see a physician, but that we should never trust man over God. Studies estimate that almost 100,000 "...people die in any given year from medical errors that occur in hospitals,"[21] identify "...a substantial amount of medical error"[22] in hospitals, and state that medication reactions account for around 100,000 deaths per year.[23] (See excerpts with the last two sources.) Now, indeed, is the time to go to the Scriptures for God's wisdom that can benefit *"all our flesh,"* (Pro 4:20-22). The three previously quoted studies are from the years 1998 and 2000. A Johns Hopkins study from 2016 adds these numbers together. It is titled: "Medical error—the third leading cause of death in the US."

In Proverbs 17:22, it also is stated that a medicine *"doeth good."* In the medical literature it is stated medication reactions cause about 100,000 deaths per year. These are reactions to *properly* prescribed, dosed, and administered pharmaceutical drugs. What God calls medicine and what man calls medicine are many times two different things; and we have been conditioned to esteem medicine that aligns with the Bible as inferior to what man calls medicine. (Do not stop taking your medications without speaking to your doctor, who understands the withdrawal reactions that can occur.) Yes, our medicines, or more accurately, pharmaceutical drugs, are made from taking things found in nature; the truth remains that man cannot improve upon God's designs. (This is not to say man cannot make *"witty inventions,"* Pro 8:12.) There is also a difference between processing something found in nature *while trying to preserve its integrity,* versus making an artificial version of it that can be patented. I am *all for* methods of preparing medicine that preserve the original design, but that is not usually man's motivation when making drugs. It is usually to make a chemically similar version for the purpose of securing a patent; it's about profit, not health. I'm not saying that it's wrong to take these medications or that they should never be used, but we should not ignore God's words while blindly following man's.

If I or someone in my family was in a serious accident or experienced some type of trauma, I will be going to the hospital as quickly as possible. I greatly value knowledgeable physicians. A few years ago my brother-in-law's life was saved by a talented surgeon. God deserves the credit though, especially since his doctor said it was a miracle that he was alive before he even had the surgery. I highly esteem good doctors and their medical knowledge. However, I am not willing to ignore what the Bible states: if a doctor says something that contradicts the Bible, they are simply wrong. I will trust what God has declared over anything that man says. Most do not know that medicine changed greatly when the Rockefellers and Carnegies gave enormous sums of money to medical schools and then controlled what kind of medicine was taught. Their teaching emphasized treating with drugs; this has changed what many considered to be medicine.

We have an entire medical system largely opposed to the truth of God's Word: that He heals, not man. This system has also rejected God in favor of evolution, meaning many of its premises are fundamentally flawed—if they are based on this falsely-so-called science. God created our bodies to be able to heal; working with the body to strengthen it to promote healing is not how modern medicine usually works. Chronic conditions are almost never healed by modern medicine.

As already stated, in an emergency, I'm heading to the hospital. If I break a bone I will let a knowledgeable doctor **who has**

studied God's creation put those bones where they need to be and then **let my body heal itself**, as God designed. If I'm bleeding to death, I'm going to want medical care that will keep the life of my flesh where it should be. If I were to be diagnosed with a chronic condition, I will find a doctor that utilizes medicine that aligns with God's Word for that situation as well.

"Strive to preserve your health, and in this you will the better succeed in proportion as you keep clear of the physicians."—Leonardo Da Vinci

"Until someone can figure out how to make substantial profits by selling such things as sunshine, fresh air, pure water, natural food and herbs, exercise, positive thinking, and self-sufficiency, you're going to have to be content with anecdotal evidence, and there is plenty of that."[24]

"With an immune system supported by clean water, exercise, adequate sleep, decent living conditions, reasonably good food, and a 25-hydroxy vitamin D level...few will be at risk of dying from an infection. In fact, most won't get sick at all. Health, after all, does not come through a needle."[25]

The human body has an amazing capacity to heal—when it is given what it needs to accomplish this. This starts with a child's first food: milk. As I read through accounts of children having vaccine reactions, a common precursor was some kind of digestive issue and a change of the child's formula was noted. As much as nursing is encouraged today, avoiding vaccines is much more important to keep our children healthy. I prefer my child to be unvaccinated and formula-fed rather than vaccinated with the perfect diet. The reality remains that formula-fed children are more likely to have health issues which lead to doctor's visits and medication prescriptions.

"Through a persistent, pervasive hypnosis from pediatricians, news outlets, print ads, web content, government press, TV programs, and magazines, parents are being taught that it is normal to have a child who can't talk, poop, walk, eat, digest food, grow, attend school, or perhaps even breathe without the benefit of a litany of medications and vaccines. Worse, we are conditioned to believe that children who *don't* join in are somehow *not* normal, or—perhaps more inexplicable—that they are a danger to children who do submit to the dozens of vaccines, antibiotics, and drugs. It is arguably the other way around, but that's another conversation. In the same time frame in which giving babies and children dozens of drugs and vaccines became normal, our children have gotten sicker, not healthier. It cannot be claimed that this style of healthcare works. ...nutrition status is what drives a child's ability to fight infections."[26]

"My kids are fine. Yours are another question. ... Un-vaccinated children get sick as they are supposed to by nature, and they get over their illnesses far quicker than vaccinated children, and develop a natural lifetime immunity to illness and a strong immune system. ... Seeing is believing, and my sole motivation is to show you the truth so your children can heal. I have seen how healthy my children are, and how unhealthy all other vaccinated children are; and the gap grows wider each year. You need to get yourself and your children off all vaccines immediately, clean up their diets, give them a lot of good quality water and foods, and watch the magic happen. If they've had a bunch of vaccines they will most likely go through a detoxification period. But in most cases the body can bounce back in a short period of time (granted they haven't been too overly poisoned by vaccines already), especially if you consult the services of a local Naturopath, Chiropractor or other natural health practitioner to aid in the healing process. ... No matter how much you've been trained, or better said, brainwashed, to believe in the medical religion...[the] fact of the matter is that it is running you and your family into a deep, dark and deadly ditch that gets worse each year. Take the steps necessary to learn something new. Open your mind to the possibility that you have been lied to for many years about the nature of your body and your environment; that you are not the frail and fragile biological instrument that falls apart every time someone sneezes in your direction. Your body and immune system is a mastery of perfection, and the emperors are truly not wearing any clothes. It's time you too let them know you know."[27]

"The false theories of...cures have made man into a mental beggar. ... So far the masses have trusted their health and life to a profession that has failed to make good."[28]

"The most important step in healing from man-made disease is realizing something you may have already begun to do: your or your children's health is your responsibility. Completely. Totally. You have more time to invest in understanding these illnesses. You have more at stake than any doctor or behavioral therapist. Although they are important partners and may have more knowledge than you do, your desire to get better will eclipse nearly anything someone else can offer you."[29]

A strong immune system is the best way to avoid the perceived need for vaccines, and the occasional genuine need for medicine that will be met with prescription drugs by the typical physician. If we feed our children properly, they are much less likely to become ill.

"Today no one believes in spontaneous generation. The remnant of this belief was annihilated through the discovery of germs as the cause for fermentation. ... No one bothered to note that for a germ to exist it has to have a house to live in—a 'house' which is created by the toxic and pathological pollution of the body, by systemic poisoning which creates the proper germ culture."[30]

When they do get sick, there are non-toxic ways to help them.

Psalm 104:14 **He causeth the grass to grow for the cattle, and herb for the service of man: that he may bring forth food out of the earth;**

"I have yet to see a single case of cold or flu, in either my personal or professional life, that does any better with a prescription of antibiotics. I suggest my patients administer homeopathic remedies for cold and flu symptoms, and these work far better than antibiotics could ever dream of working. Homeopathy is also non-toxic, whereas antibiotics are highly toxic to a child's developing bone marrow."[31]

One of the most important nutrients to keep man well is vitamin C. It is vital to keep us healthy; it helps the body fight off a host of illnesses, and also acts as a cure for many poisons, as demonstrated in Thomas Levy's book *Curing the Incurable: Vitamin C, Infectious Diseases, and Toxins.*

"To date, there are NO infectious diseases that have ever been found in which vitamin C administration can be considered dangerous or inappropriate."[32]

Levy believes that the rare ability for some humans to produce their own vitamin C can account for some of the "...anecdotal reports that one occasionally hears about a certain individual living to 100, while smoking and drinking every day of his adult life."[33] Such is life: "*...time and chance happeneth to them all*" (Ecc 9:11).

"Doctors have used vitamin C effectively in cases of measles, scarlet fever, poliomyelitis, pertussis, opiate withdrawal, coronary artery disease, cancer, life-threatening sepsis, encephalitis, pneumonia, and other conditions. Unfortunately, their work has been forgotten or actively denied."[34]

Elderberry syrup is also an effective way to help our bodies fight illness.

"The elderberry extract inhibited Human Influenza A (H1N1) infection in vitro... The Direct Binding Assay established that flavonoids from the elderberry extract bind to H1N1 virions and, when bound, block the ability of the viruses to infect host cells."[35]

"We investigated the efficacy and safety of oral elderberry syrup for treating influenza A and B infections. Sixty patients (aged 18-54 years) suffering from influenza-like symptoms for 48 h or less were enrolled in this randomized, double-blind, placebo-controlled study during the influenza season of 1999-2000 in Norway. Patients received 15 ml of elderberry or placebo syrup four times a day for 5 days, and recorded their symptoms using a visual analogue scale. Symptoms were relieved on average 4 days earlier and use of rescue medication was significantly less in those receiving elderberry extract compared with placebo. Elderberry extract seems to offer an efficient, safe and cost-effective treatment for influenza."[36]

In Suzanne Humphrey's book *Dissolving Illusions: Disease, Vaccines, and the Forgotten History,* she has a chapter titled "Lost Remedies." In this chapter, we read about natural, non-toxic remedies such as cinnamon, jicama, fresh juice, apple cider vinegar, cod liver oil, and silver, along with accounts of them being used throughout history as medicine.

"I see in your issue of last week that a surgeon in India has tried the injection of chloral in cholera, and lost 60 per cent of his patients. I have tried oil of cinnamon in an outbreak on board an Indian emigrant ship, and every case recovered. I have no doubt that the above medicine is a specific for cholera."[37]

"In 1899 Dr. C. G. Grant observed that cinnamon protected from malaria. He recommended it for intestinal problems, typhoid and influenza."[38]

"In the 1890s, General Rivera of the Mexican Army was faced with the illness of typhus fever in many of his men. A local woman administered a remedy to the soldiers with surprisingly good results. The remedy was a preparation of jicama, a sweet root vegetable native to Mexico and Central and South America. This lost hero of history used a root vegetable to treat more than 4,000 cases of pneumonia or typhus for free. Amazingly, every person was cured, and there were no reported deaths."[39]

She also tells how fresh vegetable juices were used to cure tuberculosis; smallpox was treated with apple cider vinegar; cod liver oil was used to treat puerperal fever; and colloidal silver helped to contain influenza outbreaks during the 1918 pandemic.

If you are not persuaded to forgo all vaccines for your children, or if your livelihood depends upon you receiving vaccines, I want to give you my best advice. First, understand that most pediatricians do not know much of this information. My daughter's former pediatrician had never heard that infectious diseases largely disappeared before mass immunization programs were implemented, even though it is stated in *Pediatrics,* the official journal of the American Academy of Pediatrics, that "...vaccination does not account for the impressive declines in mortality seen in the first half of the [20th] century."[40] I referenced this study already, but it is an important fact for us to remember. My daughter's current physician was greatly surprised that my child's previous doctor had never heard this.

First, find a doctor that understands your concerns, and respects your divine and sacred right to choose what is best for your child. There are delayed vaccination schedules you can follow such as the one promoted by Paul Thomas, M.D., from his book *The Vaccine Friendly Plan,* and omit the aborted fetal cell line vaccines, or get one of the ethical alternatives available (a source is listed in the additional resources). You can also do your own research and make your own schedule; many resources are available to help.

It is an undeniable fact that susceptible children will be maimed and killed because of vaccines. Yet, all children are on the same vaccination schedule without knowing who is in danger. To inject every child, without giving complete and honest informed consent to his or her parents showing the known risks, and without knowing who can take them safely and who cannot, should be considered criminal. It is imperative to find a pediatrician that understands these issues and treats your child as an individual. If you are left with a doctor who sees your child as a creature of the state that belongs on a vaccination assembly line, you must arm yourself with facts. After my experience, I greatly encourage you to find a doctor that will not contend with you and make you feel like a terrible parent every time they see your child. You might be better off with a family physician rather than a pediatrician.

If you choose to get some vaccinations for your children, the medical literature indicates they should not be given acetaminophen (Tylenol®) around the time they are given a vaccination. Acetaminophen reduces glutathione,[41] sometimes called the "master" antioxidant, which is very important regarding detoxification:

"Glutathione…is the most abundant intracellular antioxidant thiol and is central to redox defense during oxidative stress."[42]

Vaccines contain many harmful ingredients that we should not inject into any human, let alone pregnant women, helpless newborns, or slightly older infant children. Common sense would indicate it is unwise to give our children vaccines with toxic ingredients, along with a pain-killer *that inhibits their ability to handle those toxins.*

Acetaminophen use after vaccination has even been associated with autistic disorders.[43]

"Vaccine ingredients include known toxicants and carcinogens such as thimerosal (a mercury derivative), aluminum phosphate, formaldehyde (for which the Poisons Information Centre in Australia claims there is *no* acceptable safe amount that can be injected into a living human body), and phenoxyethanol (commonly known as antifreeze). Some of these ingredients are gastrointestinal toxicants, liver toxicants, respiratory toxicants, neurotoxicants, cardiovascular and blood toxicants, reproductive toxicants, and developmental toxicants, to name a few of the known dangers. Chemical ranking systems rate many vaccine ingredients among the most hazardous substances, and they are heavily regulated. Even microscopic doses of some of these ingredients are known to be able to cause serious injury."[44]

Injecting ourselves with carcinogens, multiple weakened viruses, and then giving a pain reliever that affects the body's detoxification processes might not be the best strategy for overall health or longevity. Especially since three of the four factors required for a virus *to cause cancer* also follow vaccination.

"An estimated 15 percent of all human cancers worldwide may be attributed to viruses, representing a significant portion of the global cancer burden. Both DNA and RNA viruses have been shown to be capable of causing cancer in humans. ... The path from viral infection to tumorgenesis is slow and inefficient; only a minority of infected individuals progress to cancer, usually years or even decades after primary infection. Virus infection also is generally not sufficient for cancer, and additional events and host factors, such as immunosuppression, somatic mutations, genetic predisposition, and exposure to carcinogens must also play a role."[45]

"God created a perfect reproducing organism without flaw. This genetic material [foreign RNA and DNA genetic molecules] interferes with that very principle. When the cell cannot reproduce normally, it is damaged, and acts to produce a damaged organism. This can even cause the cell to divide wildly, with toxic imperfections and out of control, which is the fundamental basis of cancer. ... Diet and environmental toxins are also a piece of the cancer puzzle, but they shy next to this vaccination component. ... You will also be hard pressed to find a case of cancer in an unvaccinated person, especially an unvaccinated child."[46]

Rather than giving children something that hinders their body's ability to handle toxins and viruses, give them something that helps their bodies rid of them safely—vitamin C. Vaccination "...is known to result in increased utilization of vitamin C..."[47] which has already been mentioned; additionally low levels of vitamin C leaves one with "...an increased risk of adverse reactions to vaccination."[48]

"No negative outcomes following a vaccination will ever occur if no vaccination is ever given. However, if you must receive a vaccination, the toxicity is greatly lessened and the desired immune response definitely enhanced by giving generous doses of vitamin C before and after the vaccination."[49]

With the help of a doctor educated in the importance of vitamin C, find good quality **liposomal** vitamin C and follow your doctor's instructions to increase your child's vitamin C levels leading up to, and also after, their vaccination. (Look for third-party testing; the supplement industry is notoriously unregulated.) Finding a doctor that can administer vitamin C intramuscularly (or intravenously for adults) would be the most effective way to increase vitamin C levels; but oral liposomal vitamin C products offer a similar effect, without any needles or appointments. Products are also available specifically designed to help protect your child from the harmful effects of vaccination, which you can purchase online.

Later, we will read an excerpt that makes a statement about not injecting vaccines because they bypass the barriers that God has in place for a reason; and I've made a similar comment about injecting metals and diseases. Some might say this logic should apply to injected vitamin C as well. I would answer this by pointing out the great difference between administering what God specifically called medicine, and something that God specifically said to avoid (disease), that He never called medicine (metals), and that does not accomplish good things in our bodies (toxins).

I also feel compelled to mention other negative consequences associated with acetaminophen use:

"Recent epidemiological observations suggest that acetaminophen (paracetamol) may contribute to asthma morbidity. Impaired endogenous antioxidant defences [sic] may have a role in the pathogenesis of a number of inflammatory pulmonary diseases, including asthma."[50]

Please find a safer pain-reliever/fever-reducer for your child, or use these over-the-counter drugs sparingly. Even when children are teething, nontoxic methods have also been used to ease their discomfort, and their irritability.

"While I could not explain all the features of teething, a consideration of the irritability in particular made the use of vitamin C an attractive proposition. I had no trouble finding [a child] to 'experiment' with. She was the daughter of a young trained nurse. There was every reason to believe the mother when she said that the diet had been well and truly supplemented with more than adequate amounts of vitamin C. An injection worked miraculously. Within thirty minutes the irritability ceased. There was a remarkable personality change. What had been a whining, crying infant, incapable of either accepting or giving affection, became a normal, loving child. ... There is little doubt that for some reason or other some infants utilize increased amounts of vitamin C while teething."[51]

(This could further explain the connection between poliomyelitis and teething: children with lower vitamin C levels could be more likely to develop symptoms of toxicity after exposure to harmful medicinal metals.)

Finally, I want to discuss children who should not receive a vaccine because they are medically predisposed to have a negative reaction. If a family member has experienced a reaction, related children are much more likely to do the same. If a child has already had a reaction, they are likely to have another, *worse* reaction. I want to point out that vaccine reactions are much more common than parents realize; what a physician calls "normal" could actually be a reaction. You've already read this next quote in this guided tour of the rabbit hole that is vaccine research, but I use it again to show how expected responses to vaccination could very well be *reactions*.

"When the baby reacts to a DPT shot with a 'slight fever and fussiness' or 'drowsiness' for a few days, this may be, and often is, a case of encephalitis which is capable of causing quite severe neurologic consequences."[52]

"...when parents are asked specifically to observe any change in their baby's behavior or physical condition after a shot, as was done by Barkin and Pichichero in Colorado in 1979, a very high proportion report unusual symptoms. In their study, <u>only 7 percent</u> of the parents reported no reaction at all; 27 percent reported mild reactions; 59 percent reported moderate reaction; and 7 percent reported severe reactions."[53]

It's amazing how many children have had a reaction to vaccines, only to be given more of the same shots by their unconcerned pediatrician two months later. So many families have been ruined because of this carelessness and ignorance. A family history of allergies, or colic, is another marker that a child could be more at risk than others.

"Some children are apparently born with an allergic sensitivity to milk in the form of a tendency to colic, and they are particularly at risk from vaccine-induced encephalitis. Others are not allergic from birth but become allergic after the shots. …encephalitis, especially from vaccination, can give rise to an allergic state, while conversely the existence of an allergic state predisposes to the development of encephalitis after vaccination. …children with allergies may overreact to the pertussis vaccine."[54]

"Although the cause of colic has never been scientifically defined, it may have an allergic basis, William J. Cates, M.D., an allergist in Rockville, Maryland, maintains: 'I have never seen a case of colic that wasn't due to allergies.'"[55]

Children born prematurely or who are underweight are also more likely to have a reaction.

"The nervous system and brain develop most in the last three months of pregnancy, and can be harmed by a premature delivery."[56]

Some babies even "…experience signs of cerebral irritation due to complications of pregnancy or difficulty in delivery. Such signs may include encephalitic screaming, opisthotonic posturing (the baby bends his head and heels backwards and arches his back), or unusually restless and irritable behavior. Whatever the cause, such signs should be reported to a physician. England considers a history of 'cerebral irritation in the neonatal period' to be a contraindication [to vaccination]. Reports from parents of vaccine-damaged children seem to support this."[57] Archie Kalokerinos gave children exhibiting signs of cerebral irritation vitamin C injections and they recovered quickly. This irritation came to be considered "…the first new symptom of <u>infantile scurvy</u> recognized…"[58] at his practice.

"'It appears reasonable, as has been done previously, to recommend that great care be exercised with every form of vaccination…' The child who is congenitally vulnerable, had a difficult birth, or had poor health in early life, is a likely candidate for vaccine damage."[59]

A family history of seizures, neurological disease, any chronic condition, and even a recent illness such as a cold, have all been considered contraindications.

"Most pediatricians and family doctors endorse and follow the CDCs recommended schedule. However, there are well-respected physicians who depart from this regiment and support vaccination choice. While they do not agree on every point, these physicians suggest that parents and individuals consider the following, to guide their vaccination decisions:

1. A family history of autoimmune, chronic, neurological, or developmental disorders, including adverse reaction to vaccines.
2. The child's medical history, including recent illness, and fevers, and use of prescription medications.
3. The number of vaccines received at a single visit, with special attention to live virus, aluminum- and mercury-containing, and combination vaccines.
4. Any indication of an adverse reaction or developmental delay is a reason to stop or slow down vaccination.
5. The use of vitamins, herbal medicine, and homeopathic remedies before, after, or instead of vaccination."[60]

Some authors even suggest to take even more factors into account:

"[1]Parents' malnutrition and use of drugs and alcohol, [2] Mother's weak nutritional statue, [3] Bottle feeding, [4] High intake of milk products, [5] High reliance on commercially processed foods, [6] High intake of sugared snacks and juices, [7] A lack of fresh vegetables and whole grains, [8] Insufficient good-quality protein foods."[61]

Archie Kalokerinos also wrote about his experience with the Australian Aboriginal population, which had a very high risk of vaccine adverse reactions.

"A health team would sweep into the area, line up all the aboriginal babies and infants and immunize them. There would be no examination, no taking of case histories, no checking on dietary deficiencies. Most infants would have colds. No wonder they died. Some would die within hours from acute vitamin C deficiency precipitated by the immunization. Others would suffer immunological insults and die later from 'pneumonia,' 'gastroenteritis,' or 'malnutrition.' If some babies and infants survived, they would be lined up again within a month for another immunization. If some managed to survive even this, they would be lined up again. Then there would be booster shots, shots for measles, polio, and even T.B. Little wonder they died. The wonder is that any survived. ... We were actually killing infants through lack of understanding."[62]

Psalm 147:5 **Great is our Lord, and of great power: his understanding is infinite.**

Remember: if you choose to receive a vaccine, none of them are *absolutely* safe.

"The Surgeon General of the United States, Leonard Scheele, pointed out to the annual AMA convention in 1955 that 'No batch of vaccine can be proven safe before it is given to children.' James R. Shannon of the National Institute of Health declared that 'The only safe vaccine is a vaccine that is never used.'"[63]

"By definition, vaccines make a healthy child sick."[64]

Keep your children healthy, and you need not fear infectious diseases, "vaccine-preventable" or not.

"Persons who are exposed to the contagion of small-pox when in a debilitated condition are liable to take the disease, while those who are strong and robust may escape entirely, or have it lightly; and the same is true of all other diseases."[65]

"For the past fifty years I have been demonstrating that the use of natural nutritional treatment is and must be the most effective form of medicine and that when it is not used and the profession depends solely on the use of toxic drugs the results are abysmal."[66]

Bloodletting was used for thousands of years even though there "...has never been any scientific evidence to support [its] efficacy..."[67] When we consider bloodletting, inoculation, and vaccination, what man calls medicine has a track record of error that spans thousands of years. While the Bible has the opposite: thousands of years of demonstrable scientific accuracy.

The implications of these last few chapters are horrifying. While most medical professionals truly believe vaccines are the miracle they are claimed to be, many in the medical profession and government agencies realize these products are harmful and will not submit their own children to the mandates they impose on others (Mat 23:4). Vaccination brings in profits, "...but also makes splendid and prospective patients for the future."[68] God warned us there would be those who would oppress others for their own gain; how much do you trust God's word?

Part III

Acts 26:24 **And as he thus spake for himself, Festus said with a loud voice, Paul, thou art beside thyself; much learning doth make thee mad.** 25 **But he said, I am not mad, most noble Festus; but speak forth the words of truth and soberness.**

Isaiah 44:24 **Thus saith the LORD, thy redeemer, and he that formed thee from the womb, I am the LORD that maketh all things; that stretcheth forth the heavens alone; that spreadeth abroad the earth by myself;** 25 **That frustrateth the tokens of the liars, and maketh diviners mad; that turneth wise men backward, and maketh their knowledge foolish;**

"Established beliefs are well nigh invulnerable because they are accorded infallibility by the masses who are educated to believe that they will be damned for thinking, and because of this, few will tolerate opposition of any nature to anything they have been educated to believe. People who have their thinking done for them are always intolerant."

J.H. Tilden, M.D.

Part III

Chapter 8

Carry Tales to Shed Blood

Ephesians 6:12 **For we wrestle not against flesh and blood, but against principalities, against powers, against the rulers of the darkness of this world, against spiritual wickedness in high places.**

Romans 3:15 **Their feet are swift to shed blood:** 16 **Destruction and misery are in their ways:** 17 **And the way of peace have they not known:** 18 **There is no fear of God before their eyes.**

God has told us in His Word to "*Prove all things; hold fast that which is good*" (1Th 5:21) *for our good*. We live in a world where genocides happen, and it is legal to kill unborn babies. This world is run by the god of this world (2Co 4:4). It is full of wickedness because of men who love "*darkness rather than light*" (Jhn 3:19). God told us that there would be men who would lie and kill to make money. God also warned us that this entire world is against Him and His followers.

The second Psalm tells us that the kings of this earth unite against the LORD and His anointed, and warns them to *be wise* (Psa 2:2-10). Ezekiel 22:9 states that men carried tales to shed blood, and verse 12 says they have taken gifts to shed blood; they had forgotten God. Verse 27 also tells us of those who are like wolves ravening their prey by shedding blood, destroying souls, and getting dishonest gain. Habakkuk describes how the wicked make other men as fish of the sea, catching them in their nets, and by them, their portion is fat and their meat plenteous (Hab 1:13-16). Then the Scripture reasons: "*Shall they therefore empty their net, and not spare continually to slay the nations?*" (Hab 1:17). Jeremiah warned the Israelites not to trust their brethren who had previously *dealt treacherously* with them, saying "*...believe them not, though they speak fair words unto thee*" (Jer 12:6). Can we not heed the same warning and recognize the terrible things done in the name of vaccine research and "the greater good" will never bring us health, though many fair words are used?

Ezekiel 22:25 **There is a conspiracy of her prophets in the midst thereof, like a roaring lion ravening the prey; they have devoured souls; they have taken the treasure and precious things; they have made her many widows in the midst thereof.**

Matthew 23:4 **For they bind heavy burdens and grievous to be borne, and lay them on men's shoulders; but they themselves will not move them with one of their fingers.**

Psalm 64:2 **Hide me from the secret counsel of the wicked; from the insurrection of the workers of iniquity: ... 5 They encourage themselves in an evil matter: they commune of laying snares privily; they say, Who shall see them?**

Habakkuk 3:14b ...**their rejoicing was as to devour the poor secretly.**

The last verse was penned thousands of years ago, but it is a prophetic passage; this is describing wicked men *today*. Guess who are the most likely to receive vaccines?

"We found that children of less educated mothers and children in Hispanic and non-Hispanic Black families with low income-to-poverty ratios were more likely to have completed the 4:3:1:3 [immunization] series."[1]

"Predictors of SIDS include individual characteristics (male sex, low birth weight, multiple births, and black race), maternal characteristics (young age, low education, and cigarette smoking), and low family income. Rates in blacks have consistently been reported to be higher than those in whites; however, in one analysis, this difference disappeared after controlling for maternal education and family income."[2]

"A majority, perhaps as many as 75 percent, of abortion clinics are in areas with high minority populations. Abortion apologists will say this is because they want to serve the poor. But you don't serve the poor by taking their money to terminate their children."—Dr. Alveda King

"...Abortion has done what the Klan only dreamed of."—Dr. Alveda King

The Bible tells us the wicked rejoice to devour the poor secretly. Women of lower socioeconomic status are more likely to get an abortion[3] and the less educated with lower incomes are more likely to complete immunization series, which we've seen are associated with higher infant mortality rates. While it is presented as foolish not to vaccinate, the actual demographic that does not vaccinate is the higher educated[4]; the most informed are the least likely to vaccinate their children.[5] I believe abortion and vaccination are true examples of systemic racism in America; they are forms of oppression whose yoke many have been tricked into putting on their own necks. Under the noble guises of "choice," "women's rights," and "health," children are harmed.

Those who have their thinking done for them are always intolerant; and those who believe abortion is good have certainly had their thinking done for them, and are intolerant of those who refuse to accept the killing of innocent children as simply another choice. A holocaust is taking place behind closed doors that most of us have been fooled to support. Instead of drums to drown out the cries of a babe offered to Molech, these things happen out of public view while men and women scream "choice" to drown out the injustice of taking away *every choice* that will ever be made by another human being. Their voices are "full of sound and fury, signifying nothing." **Everyone believes women should have a choice**. Those against abortion believe that deciding whether or not to kill your own child is not part of the decision-making process; pregnancy is the result of a choice the woman already made.

Psalm 119:110 **The wicked have laid a snare for me: yet I erred not from thy precepts.**

Psalm 119:98 **Thou through thy commandments hast made me wiser than mine enemies: for they are ever with me.**

God warned us to be wise and prove all things for a reason. God's laws keep us safe from the lies of this world. According to man's wisdom, it is acceptable, even good, to kill unborn children; and this practice was necessary to make vaccines to keep newborn children healthy. Do you truly believe that those who kill small, innocent, unborn children care about the lives of your slightly bigger, innocent children, that they couldn't convince you to kill before they were born?

God warned us that men would be without natural affection (Rom 1:31; 2Ti 3:3). We know that a mother can be convinced to kill her own child before he or she is born; is it such a stretch that some people who are removed from the child, and who would benefit financially from its death or illness, would do so? God has told us our fight is against *spiritual wickedness in high places*. Be wary of the fair words of this world. Jesus warned His followers to be "*...wise as serpents, and harmless as doves*" (Mat 10:16).

If you speak plain Biblical truth and this world doesn't like it, you might be doing something right. Labels don't matter, truth does. Jesus was called "*beside himself*" (Mar 3:21), and was put to death for telling the truth. He was also called a blasphemer (Jhn 10:33), devil-possessed (Mat 12:24), gluttonous, and a winebibber (Luk 7:34), none of which was true. What was true was that He was God in the flesh. He pointed out the sinful condition of man's heart, and the world hated Him for it. God's truth will never be popular, but the lies of this world will be. As long as Christians fail to follow God's commands to prove all things and to hear a matter before we answer it, we will be fooled right along with the rest of the world, *to our hurt*.

"[Those] who bow obsequiously to opinion, however dogmatically or arbitrarily enforced, are the enemies of mankind. They almost justify oppression by their willingness to accept its yoke. The impulse of the period, we regret to say, has been to devolve all upon the government, and curtail the freedom of the individual, and to subject the private to the public conscience, as is done in the religious establishment of the Old World."[7]

"When any class of population appears to be willing to be fooled or enslaved, it seems a fearful task to show them anything better."[8]

God warned us about many things, one of which is that men would trust in oppression. Why would men do this? Because they have been fooled into trusting things that harm them. Oppression makes a wise man mad (Ecc 7:7). If we are simple (uninformed—or even fooled), we will trust in oppression; if we are wise (informed), oppression will anger us. We have been fooled to believe vaccines bring health when, in reality, they are causing harm. This world *oppresses* the simple. We are not to marvel at the oppression of the poor (Ecc 5:8); this is the way of the world. We are supposed to do good unto all men (Gal 6:10). God's Spirit has regarded every work, and the oppression of the people of this world will not go unanswered when Christ returns (Psa 146:5-10).

We must trust and obey God's Word to avoid the snares of those who would devour secretly. God's Word gives wisdom (Psa 19:7), and understanding unto the simple (Psa 119:130). We are not to trust in man's oppression (Psa 62:10), nor in the fair words men use to persuade us to ignore God's laws.

Proverbs 14:15 **The simple <u>believeth every word</u>: but the prudent man looketh well to his going.**

Psalm 119:134 **Deliver me from the oppression of man: <u>so will I keep thy precepts</u>.**

Psalm 116:6 tells us: *"The LORD preserveth the simple: I was brought low, and he helped me."* The Lord has mercy on those who are unknowingly oppressed. Fellow Christian, you can no longer claim ignorance. In Isaiah it is written:

Isaiah 30:12 **Because ye despise this word, and <u>trust in oppression</u> and perverseness, and <u>stay thereon</u>:** 13 **Therefore this iniquity shall be to you as a breach ready to fall, swelling out in a high wall, whose <u>breaking cometh suddenly at an instant</u>.**

You now have a choice between trusting in the oppression of man, and trusting in God's word. What will you do?

Deuteronomy 11:26 **Behold, I set before you this day a blessing and a curse;**

Deuteronomy 30:15 **See, I have set before thee this day life and good, and death and evil;** ... 19 **I call heaven and earth to record this day**

against you, that I have set before you life and death, blessing and cursing: therefore choose life, that both thou and thy seed may live:

Chapter 9

Choose You This Day Whom Ye Will Serve

Hebrews 4:12 **For the word of God is quick, and powerful, and sharper than any twoedged sword, piercing even to the dividing asunder of soul and spirit, and of the joints and marrow, and is a discerner of the thoughts and intents of the heart.** 13 **Neither is there any creature that is not manifest in his sight: but all things are naked and opened unto the eyes of him with whom we have to do.**

Do you still believe that the murder of children who "would have died anyway" is the way that God has chosen for us to keep our children healthy? Here are many Bible passages in one last attempt to break through our deceitful, desperately wicked hearts (Jer 17:9). If you knowingly participate in this practice because you do not want your child to fall ill, God told the Jews that He would *bring their fears upon those who choose the things that God hates*. We should never compromise on God's laws, even if we think it is somehow in the best interest of our family.

Hosea 4:6b ...**seeing thou hast forgotten the law of thy God, I will also forget thy children.**

Isaiah 66:4 **I also will choose their delusions, and will bring their fears upon them; because when I called, none did answer; when I spake, they did not hear: but they <u>did evil</u> before mine eyes, and chose that in <u>which I delighted not</u>.**

God told His people that whatever fear had caused them to disobey His law, *He would bring to pass*. Many choose to use vaccines for their children's "health," yet it can be repeatedly demonstrated that vaccinated children are more sickly than their unvaccinated peers.

"Unvaccinated children appear to have less exposure to disease, delaying vaccination reduces exposure to disease, contracting the disease naturally leads to less disease in future, and...excessive vaccination is considered ineffective and dangerous."[1]

"...the unvaccinated are always stronger."[2]

Any "medicine" made from an aborted fetal cell line will never bring health:

- Proverbs 12:3 tells us a man shall not be established by wickedness.
- God commanded the Israelites not to follow a multitude to do evil (Exo 23:2).
- In Proverbs 10:22 we are told the blessing of the Lord has no sorrow with it.
- Ecclesiastes 8:8 states that wickedness will not deliver those that are given to it.
- The throne of iniquity has no fellowship with God (Psa 94:20-21).
- Works of violence and iniquity will not protect man (below, Isa 59:6).
- To have innocent blood upon yourself, is to procure great evil against the soul (Jer 26:15-19).
- The Bible says, "*...he that hath clean hands shall be stronger and stronger*" (Job 17:9).

Isaiah 59:6 **Their webs shall not become garments, neither shall they cover themselves with their works: their works are works of iniquity, and the act of violence is in their hands.** 7 **Their feet run to evil, and they make haste to shed innocent blood: their thoughts are thoughts of iniquity; wasting and destruction are in their paths.** 8 **The way of peace they know not; and there is no judgment in their goings: they**

have made them crooked paths: <u>whosoever goeth therein shall not know peace</u>.

The counsel of the wicked will never bring health or anything good. Yet, *"Folly is set in great dignity"* (Ecc 10:6). Satan works by removing *"the bounds"* of the people, and robbing *"their treasures"* (Isa 10:13). Satan convinces us to ignore God's Word and His laws, and steals the treasures that righteousness yields (Pro 15:6). God has preserved His Word to give us His counsel so that we will not be ignorant of Satan's devices (2Co 2:11). God warns us about trusting in wickedness over trusting Him (Psa 52). Why would Christians take the counsel of those who reject God over the very counsel of God?

Proverbs 1:25 **But ye have set at nought all my counsel, and would none of my reproof:** 26 **I also will laugh at your calamity; I will mock when your fear cometh;** 27 **When your fear cometh as desolation, and your destruction cometh as a whirlwind; when distress and anguish cometh upon you.** 28 **Then shall they call upon me, but I will not answer; they shall seek me early, but they shall not find me:** 29 **For that they hated knowledge, and did not choose the fear of the LORD:** 30 **They would none of my counsel: they despised all my reproof.** 31 **Therefore shall they eat of the fruit of their own way, and be filled with their own devices.** 32 **For the turning away of the simple shall slay them, and the prosperity of fools shall destroy them.** 33 **But whoso hearkeneth unto me shall dwell safely, and shall be quiet from fear of evil.**

(See also Psa 50:17-23; 107:11-14)

Why would we go to the world for advice that contradicts God's Word? This is what the Scriptures say about trusting in Egypt—in the Bible, Egypt is a type of the world:

Isaiah 30:1 **Woe to the rebellious children, saith the LORD, that take counsel, but not of me; and that cover with a covering, but not of my spirit, that they may <u>add sin to sin</u>:** 2 **That walk to go down into Egypt, and have not asked at my mouth; to strengthen themselves in the strength of Pharaoh, and to trust in the shadow of Egypt!** 3 **Therefore shall the strength of Pharaoh <u>be your shame</u>, and the trust in the shadow of Egypt <u>your confusion</u>.**

Christians are to trust in God *at all times* (Psa 62:4); while the Bible says trusting in man will make us cursed (Jer 17:5). The Bible also says these things about trusting in Egypt/the world/man:

- Woe to them that go to Egypt for help (Isa 31:1).
- Both Egypt and those that trust in Egypt will fall down, and fail together (Isa 31:3).
- Trusting in Egypt is like *leaning on something that will pierce you* (Isa 36:6).
- The help of man is *vain* (Psa 60:11).

Psalm 125:3 **For the rod of the wicked shall not rest upon the lot of the righteous; lest the righteous put forth their hands unto iniquity.**

Christians are to trust in the LORD, and *do good* (Psa 37:3). Believers have an obligation to live and judge righteously, and to speak the truth. We are supposed to understand that wickedness will only bring death, and warn those who have been fooled.

Proverbs 31:8 **Open thy mouth for the dumb in the cause of all such as are appointed to destruction.** 9 **Open thy mouth, judge righteously, and plead the cause of the poor and needy.**

The Scriptures say if we are deceived and trust in vanity, vanity shall be our recompense (Job 15:31); and also those "*that observe lying vanities forsake their own mercy*" (Jon 2:8). Have you been trusting your children's health to a practice that involved abortion, rather than trusting in God and His infallible word? Repent, and resolve yourself to trust every Word of God. Thank Him for His mercy, which endures forever (Psa 118:1). Entrust your children to the One who gave up His own Son for them; to the One who "*...made the worlds,*" and upholds "*all things by the word of his power...*" (Heb 1:2-3); to the only one who heals, and whose hand none can stay.

Daniel 4:35 **And all the inhabitants of the earth are reputed as nothing: and he doeth according to his will in the army of heaven, and among the inhabitants of the earth: and none can stay his hand, or say unto him, What doest thou?**

When the Jews were given the choice between releasing Jesus, a righteous man in whom no fault was found (Jhn 19:4; 6), or Barabbas, a known murderer (Mar 15:7), they chose the murderer. Now, fellow Christian, you have two options set before you—which will you choose? Will you follow the righteous ways of Jesus, or will you follow the murderous ways of this world? It's important for us to realize—God tries us (Psa 66:10; Jer 11:20; Dan 12:10) to see if we will obey His Word or this world.

Deuteronomy 13:3b …**for the LORD your God proveth you, to know whether ye love the LORD your God with all your heart and with all your soul.**

God warned us what will happen to the wicked—He will recompense their ways. His righteous words will stand forever, but the ways of this world that turn children into merchandise will be abolished (Psa 1:6; 1Co 7:31). Jeremiah 51:56 tells us "*…the LORD God of recompences shall surely requite.*" Ezekiel wrote about God pouring out His fury and recompensing for abominations (Eze 7:8). God's eye will not spare and He will not have pity: "*I will recompense thee according to thy ways and thine abominations that are in the midst of thee; and ye shall know that I am the LORD that smiteth*" (Eze 7:9). God is righteous in doing all this because man has done wickedly (Neh 9:33). If you are saved, you are safe from God's coming judgment; but if we are Christ's, why would we live according to the wicked ways of man that He will judge with wrath, fury, indignation (Hab 3:12), and destruction (Isa 42:14)?

Luke 6:46 **And why call ye me, Lord, Lord, and do not the things which I say?**

Many of the verses in this section were given to Israel, and are used here under the immediate context of trusting in something as wicked as abortion to bring life when this is in clear opposition to the Word of God. The church is not Israel, and Christians are not promised health. All of God's words are true and profitable, but we must remember which parts of the Bible are directed specifically to New Testament Christians. The Bible contains many principles that benefit this life, but following the law does not grant eternal life. The law is good if it is used lawfully (2Ti 1:8); the law shows us what sin is (Rom 7:7), and it is our schoolmaster to bring us to Christ (Gal 3:24). By the works of the law shall no flesh be justified (Gal 2:16). We can never earn our way to heaven by being good. The law shows us our sinful condition before God so that we will come to Christ and trust in His righteousness.

With that being said, the things that happened to Israel were written for our *learning*. When one reads the Old Testament, we see repeatedly the Jews ignored God's Word, were influenced by the world around them, and performed infant sacrifices right along with the rest of the world. Now, we are in the last days, and we see abortions happen every day, and even some Christians are persuaded that abortion can be used for good. New Testament Christians are specifically told that in the last days perilous times will come (2Ti 3:1-7), and we, upon whom the ends of the world are come (1Co 10:11), are to learn from Israel's experiences recorded in Scripture. We are clearly told that God is a God of recompenses, but we might not see the benefits of obeying God's Word *in this life.*

Under the current dispensation of the Church age, Christians are promised only a few things. We are told that if we will live godly in Christ, there is one thing we are guaranteed to receive: persecution (2Ti 3:12). Obeying God is not always easy, and doing the right things will not make our lives perfect. We can choose to do everything right and still suffer because we live in a fallen world. We are to take these things patiently (1Pe 2:20), and not be weary in well-doing (2Th 3:13). Obedience to God's Word will most likely benefit this life, but it might also make our time here much harder. What is certain is that Christ will judge everything a Christian does after salvation: and we will be rewarded, or we will suffer the loss of what we could have had, had we been obedient (1Co 3:9-15). The Bible says if we judge ourselves, we should not be judged (1Co 11:31); and to sow righteousness is to reap mercy (Hos 10:12).

John 14:15 **If ye love me, keep my commandments.**

John 15:10 **If ye keep my commandments, ye shall abide in my love; even as I have kept my Father's commandments, and abide in his love.**

For those who ignore God and instead believe the lies of this world, His correction is forthcoming (Psa 50:21).

Isaiah 5:21 **Woe unto them that are wise in their own eyes, and prudent in their own sight!** ... 23 **Which justify the wicked for reward, and take away the righteousness of the righteous from him!**

Proverbs 3:7 **Be not wise in thine own eyes: fear the LORD, and depart from evil.**

Chapter 10

Nehushtan

Isaiah 14:12 **How art thou fallen from heaven, O Lucifer, son of the morning! how art thou cut down to the ground, which didst weaken the nations!** 13 **For thou hast said in thine heart, I will ascend into heaven, I will exalt my throne above the stars of God: I will sit also upon the mount of the congregation, in the sides of the north:** 14 **I will ascend above the heights of the clouds; I will be like the most High.** 15 **Yet thou shalt be brought down to hell, to the sides of the pit.** … 22 **For I will rise up against them, saith the LORD of hosts, and cut off from Babylon the name, and remnant, and son, and nephew, saith the LORD.** … 24 **The LORD of hosts hath sworn, saying, Surely as I have thought, so shall it come to pass; and as I have purposed, so shall it stand:** 25 **That I will break the Assyrian in my land, and upon my mountains tread him under foot: then shall his yoke depart from off them, and his burden depart from off their shoulders.**

Satan imitates God, and from the beginning of his fall said, "*…I will be like the most High*" (Isa 14:14). He deceives by appearing to be like God, and I believe health is no exception to this rule. Because God says that *He heals,* Satan tries to deceive by being like God in this way as well. In His Word, God only calls two things medicine—a merry heart and a leaf. (Not that there can't be multiple items and multiple methods derived from these two.) In my research, I've noticed two things that man has consistently tried to use as medicine throughout time: toxic substances and material taken from diseased or dead bodies—*human and animal*. I'm not talking about consensual organ donation, I'm talking about "corpse medicine" defined as: "Corpse medicine, or medical cannibalism, is the act of using parts of deceased humans for medicinal reasons."[1]

"Across early-modern Europe the corpse was a significant element in the pharmacopeia of the medical profession and the populace. Blood, bones, fat and sweat could be ingested, prepared and smeared in order to cure a variety of ailments from epilepsy to ulcers and rheumatic pains. Often based on classical medical theory, human organic substances were thought to possess physical and spiritual virtues that no animal, mineral or plant could provide."[2]

King Charles II was treated with "…forty drops of extract of human skull"[3] (among many other things) right before he died. The touch of a "…freshly hanged man's hand was sought after to cure a variety of swellings…"[4] This mistaken understanding of using human organic substances as "medicine" has evolved into aborted fetal cell lines. False beliefs once ensconced by superstition are now held up by pseudoscience. If there is legitimate "scientific" reasoning behind these cell lines, it is still not something that should be embraced.

Toxic substances have also been used as medicine throughout history. The medicinal use of mercury and arsenic was previously mentioned in the section about polio. These metals were understood to be dangerous in large enough doses. It is now understood there could be damage from very small amounts, especially with continued exposure. The Bible tells us a "*…merry heart doeth good, like a medicine…*" (Pro 17:22), which was already mentioned. It's important to remember that God says medicine does *good* things in our bodies.

There are accounts of poison being sold as far back as ancient Rome.

"Moreover, a pestilence occurred, the greatest of any of which I have knowledge; for two thousand persons often died in Rome in a single day. Then, too, many others, not alone in the City, but throughout almost the entire empire, perished at the hands of criminals who smeared some deadly drugs on tiny needles and for pay infected people with the poison by means of these instruments. The same thing had happened before in the reign of Domitian."[5]

James Peebles, in his book about smallpox, also mentioned this account. His book is older than the book cited, and I do not know the source of his translation, but the account appears slightly differently in his book.

"Dion Casius, the Roman historical, writing of the plague which scourged Rome in the second century, relates: 'many died in another way, not only at Rome but over nearly the whole empire, through the practice of miscreants, who by means of small, poisoned needles, communicated, on being paid for it, the horrible infection so extensively that no computation could be made of the number that perished.'"[6]

The second account sounds like ancient inoculation. Who knows what was put on those needles and under what specific pretense this poison was supposed to help guard against this plague. These people were identified as criminals, so perhaps this was not a sanctioned medical practice of the day. The point is that poison has been sold as medicine throughout history because it is profitable, while giving the body nutrients and rest to heal on its own is not.

Toxic substances and material taken from deceased persons have both been used as medicine for hundreds, if not thousands, of years. We might shake our heads and wonder how people could believe these medicines would bring health, yet, *we do the same thing*. These practices have changed, but they have not disappeared. We now have the marriage of metals and aborted fetal cell lines—two things which do not align with what God's Word calls medicine—and inject them directly into the "life" of our flesh. There is no new thing under the sun (Ecc 1:9); this will never bring health.

Proverbs 16:25 **There is a way that seemeth right unto a man, but the end thereof are the ways of death.**

"In every civilized community may be found two classes with distinct and opposite interests relating to the public health—two medical schools I may call them. One holds to tradition and dogma, with two cardinal tenets in their medical creed; a drastic drug for the stomach to cure disease and a putrid decoction of decayed animal tissue for inoculation into the blood to prevent disease. The second class...places only a secondary reliance upon antidotes and drastic specifies, the main article of their creed being: Sanitary regulation, hygiene, and obedience to the immutable laws of nature. ... There is if possible more medical than theological idolatry and bigotry in the land."[7]

If Christians can be convinced to ignore God's command, *"thou shalt not kill,"* and choose to use a product made from an aborted fetal cell line, is this not idolatry? Whether it be our children, our trust in "science," our own pride, or our own private interpretation (2Pe 1:20) of half of a verse saying God uses all things for good—something or someone other than the Lord is in first place in our hearts. Something is wrong when modern medicine convinces us to disobey God's commands.

Jeremiah 17:5 **Thus saith the LORD; Cursed be the man that trusteth in man, and maketh flesh his arm, and whose heart departeth from the LORD.**

Interestingly, the symbol of medicine is the snake on a staff, which is taken from the Old Testament (Num 21). If one googles "Medicine-Snake on a Pole," the result is a pagan god associated with healing that had a serpent-entwined rod. It is admitted that the first known combination of a pole à la snake was from the Bible. (Moses was, after all, before the Greek and Roman gods. Although the connection with paganism rather than the Bible is accurate.) Two different symbols involving snakes and a staff are used by the medical profession.[8] One symbol is called the caduceus, which has two snakes entwined around a winged staff. The other symbol is called the rod of Asclepius, which has one snake wrapped around the length of a plain staff. I have read that the caduceus came from the Roman god Hermes (or the Greek version—Mercury) and the medical profession mistakenly uses it; the Greek god Aesculapius was the god of healing therefore his staff—rod of Asclepius—is supposedly the proper symbol to associate with medicine. The medically misapplied caduceus is a symbol of merchants, commerce, trade, messengers, and, interestingly, *thieves and liars*. What is even more intriguing is that the rod of Asclepius came to be associated with pagan medicine in the ancient city of Pergamos—the same city that is mentioned in the book of Revelation. Aesculapius was the Greek healing god of Pergamos. (This next quote should read and describe the rod of Asclepius instead of caduceus; it might be slightly inaccurate, but I don't blame anyone for not properly naming the imaginary staffs of fictional pagan gods.)

"The Pergamene god was the god of healing, and its emblem was a serpent. Pergamos became a great city for healing. In the courts of the temple, snakes crawled on the ground as the afflicted came from the ends of the earth to sleep in the sanctuary. They believed that healing was possible in the presence of their healing god if the sick happened to be touched by one of those slithering snakes. Much of what the people of Pergamos called medicine we could call unadulterated superstition. ... From this city, and from this background of serpent healing, came the caduceus—the medical symbol of a short staff entwined by two serpents."[9]

I've also read that *medicine and science* were worshipped in the city. For a truly fascinating connection between this ancient city, falsely-so-called medicine, and its symbol that is still associated with medicine to this day, read what Jesus said in Revelation to the angel at the church in Pergamos:

Revelation 2:12 **And to the angel of the church in Pergamos write; These things saith he which hath the sharp sword with two edges;** 13 **I know thy works, and where thou dwellest, even <u>where Satan's seat is</u>: and thou holdest fast my name, and hast not denied my faith, even in those days wherein Antipas was my faithful martyr, who was slain among you, <u>where Satan dwelleth</u>.**

Christ said that Satan's seat and the place where he dwelled were in the same city with a pagan serpent healing temple, and this is the same place where a pagan snake symbol came to represent medicine. This is interesting, to say the least. The Hippocratic Oath was originally made by swearing to pagan gods of healing, including the Pergamene god. The serpent on the pole has served its one-time healing purpose: it was a picture of Christ being made sin for us (2Co 5:21). He was lifted up just like the serpent in the wilderness (Jhn 3:14).

In the book of Numbers, the Israelites sinned by *complaining* against their spiritual leaders, and therefore against God (Num 21:5; see also Exo 16:8), and were bitten by snakes as punishment. After they repented, God gave them a way to be cured: they had to look upon a brass snake that God instructed Moses to put on a pole, and they would be healed (Num 21:6-8). This brass snake came to be worshipped hundreds of years later and was destroyed (2Ki 18:4).

I believe much suffering and evil has been brought into this world in the name of health. "For the greater good," *terrible* medical experiments have been performed on handicapped individuals, helpless babies,[10] slaves,[11] Jews during the holocaust, and other oppressed peoples all throughout time. The pharmaceutical company I. G. Farben was "...a former designer and operator of the concentration camps used in Nazi Germany."[12] Vaccines and pharmaceutical drugs have been repeatedly tested on humans deemed less valuable than others.

God says He heals, yet we have a worldly system that claims health. This medical system that *kills children*, has a snake as its emblem, has committed atrocities against Israel, and that many obey as though infallible even when it clearly contradicts God's Word—will be destroyed. I believe we have a picture of this foretold in the Old Testament.

When Christ does rule and reign, the medicine that man uses will be much different than the current system established by man. The "*leaves of the tree*" will heal the nations (Rev 22:2). Doctors who perform abortions for women's "health," and then use the children's bodies to experiment and make "medicine" that is then injected into nearly every person on this earth will be no more. I believe we were given a picture of Christ destroying this medical system with its chosen symbol of the serpent, and it was foreshadowed by Hezekiah's (who is type of Christ) destruction of the brass serpent that came to be worshipped; as an insult, he called it ***Nehushtan***.

I only say this to show an interesting connection. Christians are here to spread the gospel and make disciples, not to try to fix the systems of this world. Of course we should obey God and do right, but this world will not be set as it should be until Christ comes. I'm not claiming it is wrong to work in this healthcare system; I'm only suggesting that Christians be ***discerning*** (Pro 2:11; Pro 3:21-23), and not to let anything a doctor says persuade you to ignore God's Word.

"Injecting poisons directly into the bloodstreams of our newborns is the hallmark of bad ideas gone worse; a prime example of how we are still existing under the auspices of an antiquated medical religion, which exists today solely for its own gains it preys upon innocent children..."[13]

"There are two forces involved: the forces of good and evil. We have the God I worship versus the god of pseudoscience, the god of compulsion versus my God who gave me inalienable rights and responsibilities to protect my children. We have been asked to voluntarily offer our healthy children as living sacrifices to a false god. The God I worship demands obedience rather than sacrifice."[14]

While it is believed there are many religions, there are really only two: the pure, undefiled religion of the Bible (Jas 1:27) and the false religion of Babylon that has many different forms. The Mystery Babylon Religion is the mother of harlots (Rev 17:5): all false religions come from Babylon. Whether one worships the *"queen of heaven"* (Jer 7:18), *"Tammuz"* (Eze 8:14), a supposedly immaculately conceived child, or any of the other idols given different names in different cultures throughout the millennia, they go all the way back to Babel. One modern version of Mystery Babylon is the religion of falsely-so-called *Science*—"Mother Nature" and her "immaculate" child *creation* with its evolved creatures, who are worshipped more than the Creator (Rom 1:25). It was foretold in Genesis 3:15 that a woman would conceive the Messiah without a man because of the words *"her seed"*; this pointed to a miraculous virgin birth. This false religion of Babylon is an imitation of the true faith which came through Christ and His virgin birth (Isa 7:14; Luk 1:34-35) that made God a man to lay down His life for the sins of mankind. The best way for Satan to deceive man is by imitating God, and it has worked throughout human history. The first false religion is also the last false religion; even what many esteem as Christianity is another version of the Babylon Mystery Religion.

Revelation 18:23b ...**for thy merchants were the great men of the earth; for by thy sorceries were all nations deceived.** 24 **And in her was found the blood of prophets, and of saints, and of all that were slain upon the earth.**

Proverbs chapter two tells us that the wisdom of God's Word shall preserve us and deliver us from the strange woman—the Babylon Mystery Religion. Here's the point of this brief mystery Babylon detour:

The Bible says in her was the blood of *all* that were slain upon the earth. If the deaths of other humans are required to accomplish something "good" in your belief system, you have *a false system.* (Eugenics, survival of the fittest, or the supposedly "righteous" slaying of "infidels," "heretics," etc.) Whether you acknowledge your beliefs are religious or not, you are involved in this false religion. As the Bible prophetically says, "*come out of her*" (see below). When the Israelites disobeyed God, they were influenced by the pagans around them and began to sacrifice their own children "*unto devils*" (Psa 106:37). There is no way around the fact that killing children is wicked.

Revelation 18:4 **And I heard another voice from heaven, saying, Come out of her, my people, that ye be not partakers of her sins, and that ye receive not of her plagues.**

Ecclesiastes 7:26 **And I find more bitter than death the woman** [Mystery Babylon Religion]**, whose heart is snares and nets, and her hands as bands: whoso pleaseth God shall escape from her; but the sinner shall be taken by her.**

In the book of Daniel, all of the wise men *of Babylon* had one-tenth the wisdom of the God-fearing men blessed with wisdom: Daniel, Hananiah, Mishael, and Azariah.

Daniel 1:17 **As for these four children, God gave them knowledge and skill in all learning and wisdom: and Daniel had understanding in all visions and dreams.** ... 20 **And in all matters of wisdom and understanding, that the king enquired of them, he found them ten times better than all the magicians and astrologers that were in all his realm.**

It is no coincidence that in the dark ages, when the Word of God was deliberately kept from the people, disease was much more common—and deadly. People were without the Bible's disease-preventing instructions, like burying waste instead of throwing it in the streets, and the warning of "*woe*" to men who live right on top of one another (Isa 5:8). This ignorance of the basic laws of cleanliness caused much needless suffering.

"It was this persistent opposition to the achievements of the past that contributed immensely in causing a veil to fall over the West, plunging it into the Dark Ages (AD 476-1453). The spurious books of the Latin Vulgate opened the door for the mysterious and for the dark doctrines which had confused the thinking of the ancients."[15]

The established superstitions and traditions from the Dark Ages allowed the people to believe that "medicine" made from poisons and corpses would bring health. Still today, most still believe we must pay a doctor to give us aborted fetal cell line vaccines—a combination of poisons and material taken from the dead—for our "health." The ignorance and superstition of the Dark Ages has carried through to today in the belief that we need to pay someone to give us good health; that they have special knowledge the average person is incapable of understanding, and dark methods must be employed to make "medicine." Doctors certainly have knowledge that the layman does not; the basic tenets of human health are not part of any secret knowledge of a privileged class. In the past, living in filth made health nearly unattainable. Today, and for most people, health is within reach and maintainable, and no questionable medical practices are required. The immune system, vaccines, and the combination of the two are all shrouded in mystery, and we assume that modern doctors have all of the intricacies figured out—the reality is they don't. This aura of mystery that a class of educated people supposedly understands leaves us trusting of this medical system and makes us easier to deceive. Treating any person as though they are infallible is foolish, even those in the medical profession. One of my favorite bloggers made this statement: "I pray parents will seek their guidance from the infallible Lord instead of government experts or profiteers disguised as doctors. We must STOP looking at doctors as gods."

False Christianity is to Biblical Christianity what false medical systems are to human health. Both are full of decent yet deceived people trying to do right, and both have a few true tenets, but they are far from the truth that matters most. Few are saved in, and none are saved by the false beliefs of these systems. Health is not mysterious; it does not require a medical "high priest" or infant sacrifice. Health is a simple, uncomplicated process. When someone complicates and conceals truth and then coerces you to ignore the Bible because *they* have a "greater understanding," we have a doctrine that Christ hates (Rev 2:6): a class of people ruling over others.

"But when mankind awakens to a full realization of the truth that throughout the hoary centuries we have been buying false remedies from monopolistic pretenders, then ancient Aesculapius will be unfrocked and thrown out of business—staff, snake, and all."[16]

One of the reasons God gave the land of Canaan to Israel was to recompense the wickedness of the people that inhabited the land, including offering their children to idols (Lev 18:21-25). God told Israel that the punishment for killing an innocent child this way was to receive the harshest penalty.

Leviticus 20:1 **And the LORD spake unto Moses, saying,** 2 **Again, thou shalt say to the children of Israel, Whosoever he be of the children of Israel, or of the strangers that sojourn in Israel, that giveth any of his seed unto Molech; he shall surely be put to death: the people of the land shall stone him with stones.** 3 **And I will set my face against that man, and will cut him off from among his people; because he hath given of his seed unto Molech, to defile my sanctuary, and to profane my holy name.**

Today, we are NOT under this law, and the Scriptures do NOT condone violence against those who perform abortions. Jesus said to be harmless as doves (Mat 10:16). Born-again Christians are considered priests before God (Rev 1:6); this passage continues to say God will set His face *against the priest* that looks the other way and refuses to judge this practice that God calls an abomination (Deu 18:10-14).

Leviticus 20:4 **And if the people of the land do any ways hide their eyes from the man, when he giveth of his seed unto Molech, and kill him not:** 5 **Then I will set my face against that man, and against his family, and will cut him off, and all that go a whoring after him, to commit whoredom with Molech, from among their people.**

Notice the person committing the crime and the priest who *hid his eyes* from the practice were both told that God would set His face against them, and cut them off. The priest that did not perform any sacrifices, yet did not judge them as God commanded, was still considered to be worshipping Molech!

Exodus 23:33 **They shall not dwell in thy land, lest they make thee sin against me: for if thou serve their gods, it will surely be a snare unto thee.**

An Old Testament priest that served in God's Temple was not to allow, or ignore, the practice of infant sacrifice in any way. Born-again believers are New Testament priests (1Pe 2:5-9), and our bodies are the temple of God: the injection of a product made from an aborted fetal cell line has no place in this temple.

2 Corinthians 6:16 **And what agreement hath the temple of God with idols? for ye are the temple of the living God; as God hath said, I will dwell in them, and walk in them; and I will be their God, and they shall be my people.** 17 **Wherefore come out from among them, and be ye separate, saith the Lord, and touch not the unclean thing; and I will receive you,** 18 **And will be a Father unto you, and ye shall be my sons and daughters, saith the Lord Almighty.**

Man's religion (Gal 1:13) and man's traditions (Col 2:8) do not please God. Man might deceive by promising life (Eze 13:22), but false religious systems that require others to die will never bring life, and even peaceful religions will not lead to heaven. The only death that brought life was Christ's sacrifice on the cross—Jesus is the only way to have everlasting life. Because of sin, we live in a fallen world; man's remedies to prevent death are lies that will never bring life. The only way to eternal life is through Jesus Christ; and if you are truly born-again, *you will get a new body* (Rom 8:23).

1 Corinthians 15:50 **Now this I say, brethren, that flesh and blood cannot inherit the kingdom of God; neither doth corruption inherit incorruption.** 51 **Behold, I shew you a mystery; We shall not all sleep, but we shall all be changed,** 52 **In a moment, in the twinkling of an eye, at the last trump: for the trumpet shall sound, and the dead shall be raised incorruptible, and we shall be changed.** 53 **For this corruptible must put on incorruption, and this mortal must put on immortality.** 54 **So when this corruptible shall have put on incorruption, and this mortal shall have put on immortality, then shall be brought to pass the saying that is written, Death is swallowed up in victory.** 55 **O death, where is thy sting? O grave, where is thy victory?**

56 **The sting of death is sin; and the strength of sin is the law.** 57 **But thanks be to God, which giveth us the victory <u>through our Lord Jesus Christ</u>.** 58 **Therefore, my beloved brethren, be ye stedfast, unmoveable, always abounding in the work of the Lord, forasmuch as ye know that your labour is not in vain in the Lord.**

This is man's only hope! What good is a long life if we do not know what comes after? This is the only death God designed to benefit man. Man's efforts to extend life through death are unprofitable; it is a false attempt at a physical imitation of the spiritual truth that all men must accept to be saved:

Romans 10:9 **That if thou shalt confess with thy mouth the Lord Jesus, and shalt believe in thine heart that God hath raised him from the dead, thou shalt be saved.** 10 **For with the heart man believeth unto righteousness; and with the mouth confession is made unto salvation.**

After salvation, we must look to God's Word to see what is good, and see the way wherein we should walk (Psa 119:9; Psa 143:8; Pro 20:24).

Micah 6:7 **Will the LORD be pleased with thousands of rams, or with ten thousands of rivers of oil? shall I give my firstborn for my transgression, the fruit of my body for the sin of my soul?** 8 **He hath shewed thee, O man, what is good; and what doth the LORD require of thee, but to do justly, and to love mercy, and to walk humbly with thy God?**

Chapter 11

Days of Noah

Genesis 6:11 **The earth also was corrupt before God, and the earth was filled with violence.** 12 **And God looked upon the earth, and, behold, it was corrupt; for all flesh had corrupted his way upon the earth.**

Luke 17:26 **And as it was in the days of Noe, so shall it be also in the days of the Son of man.**

Throughout history, Satan has tried to pollute man's DNA (Gen 6:4). The Bible tells us right before the flood *all* flesh was corrupted (Gen 6:12), except Noah who was perfect in his generations (notice generations; Gen 6:9). This was partly the reason God destroyed the world, the other being that man's thoughts were only evil continually (Gen 6:5). Also note this connection between flesh being corrupted and the wickedness on earth; were men's thoughts so evil because of this corruption of their DNA? God did not intend for us to inject the DNA of murdered children into our children to promote health, and it can be demonstrated that this is a harmful practice. God also did not intend for human and animal DNA to mingle. Corrupting man's DNA is an ancient attack on God's creation, and it is still happening. There is truly no new thing under the sun (Ecc 1:9).

Even some who do not have a Biblical worldview understand that we should not mix human and animal DNA.

"We have violated ancient taboos, held by virtually every civilization in history, by mixing the blood of man with the blood of animals. Even forbidden in the Bible, somewhere."[1]

It was not until we took the blood of beasts and mixed it with human blood through an injection that two new medical terms became necessary:

"The introduction of horse blood into humans was not without problems. It heralded the introduction of two new medical concepts—*allergy* and a*naphylaxis*."[2]

The encephalitis caused by vaccination is also considered an allergic response. This has been demonstrated in animal experiments by producing "...brain inflammation in monkeys merely by injecting them repeatedly with extracts of sterile normal rabbit brain and spinal cord material. Encephalitis was now seen to be an *allergic* phenomenon, and this explains the association of allergies and autoimmune states with a prior case of encephalitis."[3] Putting animal DNA fragments in the life of our flesh, "...a medium not intended for them, [provokes] redoubtable manifestations of the gravest morbid phenomena,"[4] including encephalitis.

God has also always told His believers to abstain from ingesting blood. Before (Gen 9:4), during (Lev 3:17), and after (Act 15:20) the law. Moses wrote in the book of Deuteronomy:

Deuteronomy 12:23 **Only be sure that thou eat not the blood: for the blood is the life; and thou mayest not eat the life with the flesh.** 24 **Thou shalt not eat it; thou shalt pour it upon the earth as water.** 25 **Thou shalt not eat it; that it may go well with thee, and with thy children after thee, when thou shalt do that which is right in the sight of the LORD.**

Vaccines made from animal products can be made from animal blood.

"Animal-derived products used in vaccine manufacture can include amino acids, glycerol, detergents, gelatin, enzymes and blood."[5]

Would the God who has *always* commanded man to abstain from blood want us to inject even trace amounts of animal blood into *our* own, the life of a beast into the life of our flesh?

Leviticus 17:10 **And whatsoever man there be of the house of Israel, or of the strangers that sojourn among you, that eateth any manner of blood; I will even set my face against that soul that eateth blood, and will cut him off from among his people.** 11 **For the life of the flesh is in the blood: and I have given it to you upon the altar to make an atonement for your souls: for it is the blood that maketh an atonement for the soul.**

In Leviticus God also said in the same sentence not to "*eat any thing with the blood: neither shall ye use enchantment, nor observe times*" (Lev 19:26). In 1 Samuel 14:33, eating with the blood is considered a sin against the LORD.

Here are other problems that have arisen from using animal cell lines to develop vaccines.

"Some of the polio vaccine administered from 1955–1963 was contaminated with a virus, called simian virus 40 (SV40). The virus came from the monkey kidney cell cultures used to produce the vaccine."[6]

"The polyomavirus simian virus 40 (SV40) is a known oncogenic DNA virus which induces primary brain and bone cancers, malignant mesothelioma, and lymphomas in laboratory animals. Persuasive evidence now indicates that SV40 is causing infections in humans today and represents an emerging pathogen. A meta-analysis of molecular, pathological, and clinical data from 1,793 cancer patients indicates that there is a significant excess risk of SV40 associated with human primary brain cancers, primary bone cancers, malignant mesothelioma, and non-Hodgkin's lymphoma. … The polyomavirus simian virus 40 (SV40) is a potent DNA tumor virus, and mounting evidence suggests that it is an emergent human pathogen."[7]

"The recent emergence of bovine spongiform encephalopathy (BSE) and variant Creutzfeldt-Jakob Disease (vCJD) suggests that transmissible spongiform encephalopathies (TSEs) pose an ongoing threat to human and animal health. To avoid iatrogenic transmission of TSEs in vaccines, strategies must be developed to obviate TSE agent infectivity in cellular substrates, cell culture media components and enzymes, and excipients, and to validate the safety of these components and field vaccines efficiently."[8]

"Transmissible spongiform encephalopathy (TSE) agents have contaminated human tissue–derived medical products, human blood components, and animal vaccines. … Fortunately, no human vaccine has ever been implicated as a source of iatrogenic TSE. However, 2 animal tissue–derived vaccines have caused outbreaks of scrapie in sheep, and 2 medical products of human origin—dura mater allograft and human cadaveric pituitary hormones (no longer marketed in the United States)—have transmitted hundreds of cases of CJD; corneal grafts have transmitted a few cases as well."[9]

The last condition has never been documented to have been transmitted to humans via a contaminated vaccine, and maybe it will not happen in the future. But, it's not something I'm willing to risk with my child.

Exposing ourselves to animal DNA is changing our DNA—through vaccination.

"The medical authorities keep lying. Vaccination has been a disaster on the immune system. It actually causes a lot of illnesses. We are changing our genetic code through vaccination..."[10] Guylaine Lanctot, M.D.

"...a virus has the annoying talent of being able to splice itself into the DNA of a host cell. Taken together, the collective DNA of our race is known as the **human genome**. Continuing to add more and more lab-altered versions of unknown foreign agents into our children's blood year after year is certain to be diluting the human genome. This long-term effect within our species is an entire field that has never been studied. ... Viera [Scheibner] uses the xerox analogy: as 16 billion injections a year access the DNA of human children, it's like making a copy of a copy of a copy, etc. The human genome is being diluted. The implications of such a trend on our species could hardly be more profound."[11]

"...There is ground for believing that the injected germ proteins hybridize with the body proteins to form new tribes, half animal and half human, whose characteristics and effects cannot be predicted... Even non-toxic bacterial substances sometimes hybridize with serum albumins to form specific poisons which continue to multiply, breed and cross-breed ad infinitum, doing untold harm as its reproductivity may continue while life lasts."[12]

We have seen much evidence to show that vaccination can cause encephalitis, and this can cause serious brain damage. It is an even lesser known fact that the damage caused by encephalitis can be less extreme, yet these lesser effects can be even more insidious in our society. Encephalitis can disrupt the myelination of one's brain. Myelination first occurs in the lower parts of the brain to the higher—the parts that distinguish us from the animals.

"In the brain, myelination begins the phylogenetically older parts (those areas which humans share with lower animals) and then moves to the phylogenetically more recent parts (which distinguish humans from animals)."[13]

I have already used this next quote, but it has been known for a hundred years that vaccination can cause encephalitis, affecting the myelination process.

"The association between the post-encephalitic syndrome and demyelination or incomplete myelination of the brain seems quite secure, and the fact that encephalitis—*including that caused by vaccination*—can cause demyelination has been known since the 1920s!"[14]

We are injecting our children with foreign animal matter, and the effect of this can be encephalitis, which can prevent the *higher parts* of the *human* brain from being myelinated properly.

"…the pertussis neurotoxin, does not leave its imprint in an all-or-nothing fashion, i.e., neurological catastrophe or full recovery, but rather in varying degrees of disability, usually requiring maturation to illuminate."[15]

"No less important than the physical manifestations of the post-encephalitic syndrome are its intellectual and moral ones. Even the mildly affected individual often shows such symptoms as egotism, narcissism, ego weakness, alienation, impulsiveness, emotional lability, flat affect, anxiety, paranoia, impatience with criticism, rage, depression and suicidal impulses. Post-encephalitis individuals have fragmented thinking, and their memory is weak; hence they cannot think rationally and even lack ordinary common sense. Compensation for these perceived weaknesses often takes the form of aggression against other, hypersexuality, homosexuality, alcohol or drug abuse, and suicide. Very probably these behaviors are also direct manifestations of the underlying neurological weakness. The long-term consequences of rage, aggression, premature sexuality, and drug and alcohol abuse in this population is the ongoing epidemic of violent crime in the United States… While the vaccination program is, of course, not responsible for every instance of the above disabilities and social woes, it nonetheless makes a substantial contribution to them."[16]

"These children were often of normal or near-normal intelligence—'moral rather than mental imbeciles.'"[17]

"…five percent of prisoners have a history of seizures according to their own accounts, or have been taking anti-seizure medication. This is five to ten times higher than the prevalence of epilepsy in the population at large [this source was published in 1990]."[18] (Seizures can be an adverse reaction to vaccination.)

"Thus, wherever one looks—even in the finer points of personality development—the parallels between autism, minimal brain damage, the post-encephalitic syndrome, and vaccine damage are overwhelming. … Children with post-encephalitic syndrome lack deep feelings. 'Emotional poverty was an outstanding symptom.'"[19]

Genesis 6:5 **And GOD saw that the wickedness of man was great in the earth, and that every imagination of the thoughts of his heart was only evil continually.** … 12 **And God looked upon the earth, and, behold, it was corrupt; for all flesh had corrupted his way upon the earth.**

Matthew 24:37 **But as the days of Noe were, so shall also the coming of the Son of man be.**

2 Timothy 3:1 **This know also, that in the last days perilous times shall come. 2 For men shall be lovers of their own selves, covetous, boasters, proud, blasphemers, disobedient to parents, unthankful, unholy, 3 Without natural affection, trucebreakers, false accusers, incontinent, fierce, despisers of those that are good, 4 Traitors, heady, highminded, lovers of pleasures more than lovers of God; 5 Having a form of godliness, but denying the power thereof: from such turn away.**

Humans are displaying the basest innate wickedness, cruelty, and violence. The biggest contributor to this is the belief that there is no God. Children are educated to believe in evolution, which tells them they are animals (see Pro 23:7a); they are injected with animal DNA making their genetic makeup more like an animal; and these injections can cause brain damage that diminishes their thinking capacity—making them think and act more like animals.

What are we doing to our children? "The answer is, we do not know."[20]

"One very hostile and violent sixteen-year-old…started biting himself at age fourteen months: … He says, 'I hate the way I am. How long will I have to be this way? Is it forever?"[21]

A one-year-old child biting himself is completely unnatural; are our children's brains being systematically damaged?

"What all of these milder cases share with the more serious ones is alienation: 'These three autistic adults all conveyed the deep, unquenchable loneliness, the conviction that they were somehow incapable of forming a relationship which could satisfy them.' … For every 'autistic' who is shut away in an institution there will be a thousand alienated individuals functioning as normal taxpaying citizens."[22]

"It is not easy to discern the outlines of the incubus which the vaccination program has loosed upon us."[23]

"A major cause of the Roman Empire's decline, after 6 centuries of world dominance was its replacement of stone aqueducts by lead pipes for the transport and supply of drinking water. Roman engineers, the best in the world, turned their fellow citizens into neurological cripples. Today our own 'best and brightest' with the best intentions, achieve the same end through childhood vaccination programs yielding the modern scourges of hyperactivity, learning disabilities, autism, appetite disorders, and impulsive violence."[24]

Genesis 6:11 **The earth also was corrupt before God, and the earth was filled with violence.**

The DPT vaccine has been replaced with the admittedly safer DTaP for over twenty years, but most of the US population has received the crude and more reactive DPT shot that caused encephalitis and this minor brain damage with greater frequency than DTaP. (DPT has not been taken off the market; children in third world countries still receive this shot.) It's interesting that other countries started using the DTaP *years* before the US.

"The question remains: if Japan began using a safer vaccine in 1981, why did it take the Unites States 15 more years to license a purified one for American babies?"[25]

"Pertussis toxin...is one of the most toxic substances in nature, a true bacterial exotoxin with an enormous range of effects on the living [organism]."[26]

If vaccines can cause major brain damage, and it is undeniable that they do in certain cases, they can certainly cause lesser damage as well.

"...[Autism] belongs to a continuum of developmental brain disorder variously classified as mental retardation, cerebral palsy, and learning disabilities, and would appear to share the same causes... None of the EEGs, CT scans, or physical findings...for autism can reliably distinguish it from other developmental brain disorders."[27]

Rising autism rates also coincide with the implementation of vaccines made from aborted fetal cell lines.

"...vaccine products contain DNA fragments from fetal tissue that cannot be filtered out. Research conducted suggests that using foreign, human DNA fragments is having unintended consequences. ... Merck's **MMR II** vaccine (as well as the **chickenpox, DTaP** (Pentacel), and all **Hep-A** containing vaccines) is manufactured using human fetal cell lines and are heavily contaminated from the production process [and are] known to activate Toll-like receptor 9 (TLR9), which cause autoimmune attacks. ... **Anyone who says that the fetal DNA contaminating our vaccines is harmless either does not know anything about immunity and Toll-like receptors or they are not telling the truth. ... This is direct biological evidence that fetal DNA contaminants in vaccines *are not* in low innocuous amounts. They are a very strong proinflammatory trigger.** ... Administration of fragments of non-self human fetal DNA to a child (as in shots), could generate an immune response that would also

cross-react with the child's own DNA. Children with autistic disorder have antibodies against human DNA in their circulation that non-autistic children do not have. These antibodies may be involved in autoimmune attacks in autistic children. ... Duke University demonstrated in a recent study that significant improvements in behavior were observed when children with autism spectrum disorder were treated with their own banked autologous cord blood. This treatment **clearly shows that most children with autism are not born with it** since genetic diseases like Down syndrome or muscular fibrosis cannot be treated with autologous stem cells. Therefore, **an environmental trigger(s), introduced to the world around 1980 when autism first began to rise, must be identified and eliminated** or reduced in the environment. Strong change-point correlation exists between rising autism rates and the vaccine manufacturing switch from animal-derived cell lines for rubella vaccine to human aborted cell lines in the late 70s."[28]

The post-encephalitic syndrome causes changes in the brain which can manifest as "hypersexuality [and] homosexuality," as one of the previous excerpts stated. This next quote gives a plausible explanation for this connection:

"**Consider this:–WI-38 is the cell line from a FEMALE aborted fetus**, used to cultivate viruses used in vaccines. When you inject the DNA from a FEMALE (carrying two X chromosomes) into a MALE (who already carries one X chromosome and a weaker Y chromosome) you now have an overload of the X chromosome. *Now we have an onslaught of BOYS who think they should be GIRLS.* Do we have male DNA in vaccines? YES! –**The MRC-5 is the code given to the fetal cell line from a MALE aborted fetus,** also used to cultivate vaccine viral components. Do we have girls thinking they are boys? YES! **Is it as prominent as boys wanting to be girls? NO!** Why? Because girls have two dominant X chromosomes. When they are injected with a vaccine containing MRC-5, they aren't just getting a Y chromosome, but yet another dominant X chromosome, on top of the two they already have."[29]

Hopefully, this wide range of effects caused by vaccination will help us to be gracious, loving, and merciful to those who have had their brain's damaged because of nothing they did or chose. For example, a misbehaving child might be hyperactive through no fault of his or her own.

Man was made just a little lower than the angels (Psa 8:5), and became a living soul (Gen 2:7). He was not meant to have his DNA and brain denigrated to that of an animal; much of our societal woes come from the human brain being damaged in this way. Through becoming a born-again believer of Christ, and through God's Word and the power of the Holy Ghost which guides into all truth (Jhn 16:13), many can recover from minor brain damage that can manifest in one being a "moral imbecile." The brain has an amazing ability to heal. We can change much about how we think just by *choosing to change* our thinking, and actually *doing* it. Those whose minds have been damaged beyond repair, and for the rest of their lives have the thinking capacity of a child—will be made whole (more on this in Chapter 13).

We do so much harm in the name of "health" when vaccines are not what stopped infectious diseases in industrialized countries.

"Vaccines did not save humanity and never will. Vaccines have never been proven truly safe except for perhaps the parameters of immediate death or some specific adverse events within up to 4 weeks. Smallpox was not eradicated by vaccines as many doctors readily say it was. They say this out of conditioning rather than out of understanding the history or science. Polio virus was not responsible for the paralysis in the first part of the 20th century. Polio vaccine research, development, testing and distribution has committed atrocities upon primates and humanity. Bill Gates is not a humanitarian. Vaccines are dangerous and should never be injected into anyone for any reason. They are not the answer to infectious diseases. There are many more sustainable and benevolent solutions than vaccines."[30]—Suzanne Humphries, MD

Speaking of Bill Gates, well, I'll just let him speak for himself:

"First, we've got population. The world today has 6.8 billion people. That's headed up to about nine billion. Now, if we do a really great job on new vaccines, health care, reproductive health services, we could lower that by, perhaps, 10 or 15 percent."[31]
—Bill Gates, Innovating to Zero Ted Talk

Proverbs 30:14 **There is a generation, whose teeth are as swords, and their jaw teeth as knives, to devour the poor from off the earth, and the needy from among men.**

Bill Gates claims when health improves in a population, people have fewer children. What appears to be a sinister admission on his part is the outworking of the claim that better health leads to a lower population through fewer childbirths. So, according to Bill Gates, lives saved leads to a lower population; somehow more lives equal fewer people. I'm not sure what kind of devil's arithmetic he utilized in the algorithm that yielded this equation, but I fear skewed logic and technicalities were used to be misleading. Exactly like the narrative about two cell lines made from two abortions and no further sources were needed—there is much wickedness concealed behind this statement. I believe this claim about population reduction is also a carefully designed, technically true statement made to conceal. It is also unclear how health is defined according to Bill Gates. If the presence of "reproductive health services," such as abortion clinics, in a population is considered the "improved health" that leads to fewer children being born, this is not the kind of health we want.

Proverbs 21:6 **The getting of treasures by a lying tongue is a vanity tossed to and fro of them that seek death.**

Psalm 55:21 **The words of his mouth were smoother than butter, but war was in his heart: his words were softer than oil, yet were they drawn swords.**

Scripture also tells us that wicked men "*speak peace to their neighbours, but mischief is in their hearts. …they regard not the works of the LORD, nor the operation of his hands*" (Psa 28:3-5). Wicked men do not value human life—the work of God's hands; but they "*delight in lies*" (Psa 62:4).

Many will find it hard to believe that the Bill and Melinda Gates Foundation's vaccination campaigns are doing harm to the underprivileged around the world. The Bible clearly tells us when Christ returns, He will destroy those who rejoice to devour the poor **secretly** (Hab 3:12-14), and we cannot ignore the many verses describing those who set snares to **secretly** kill the poor. It can be clearly demonstrated that vaccines cause harm, and they did not stop any infectious diseases; yet the entire world believes vaccines have saved humankind from disease. If there is a better candidate for something that devours the poor secretly, I'm not sure what it could be. If you do not believe vaccines are the obvious vehicle for such a deception, I advise you to discover what it is you believe the wicked use to devour the poor so that you can avoid their snares.

When Bill Gates was asked during a 2015 interview if those who don't vaccinate their children anger him, he replied: "I get more mad about the deaths we're not avoiding. … I wouldn't say I get angry but I'm really impatient that we're not moving as fast as I'd like."[32] (He said this implying that vaccines save lives.) Bill Gates specifically mentions the measles vaccine, vaccines for respiratory and diarrheal diseases, which almost all children in the US will get under the current immunization schedule. It should give us pause when an elitist billionaire bent on population control supports the vaccines we give to nearly every child in the world: "*...one sinner destroyeth much good*" (Ecc 9:18).

Proverbs 26:24 **He that hateth dissembleth with his lips, and layeth up deceit within him;** 25 **When he speaketh fair, believe him not: for there are seven abominations in his heart.**

"Let's not forget - these aren't the original bugs associated with the disease they're using here, but rather manmade, lab versions of the original microbes. Harvard Medical School's Dr. Richard Moskowitz explains that the way vaccines are prepared is to make them weaker and weaker, just to the point where they don't produce an immune response. The problem is, in this form, the altered microbes can penetrate far deeper into our tissues than would naturally have been possible. There they can become latent, hidden allergens, just waiting. Like a slow virus. Then when something triggers them into action, even years later, the stowaway microorganisms can manifest themselves in virtually any location or system of the body, causing dysfunction, chaos, degenerative disease, or even death. But not from the original disease. That way no one can ever prove the vaccine was the cause of death - it's a beautiful thing."[33]

The frightening reality is that to some, the death caused secretly by vaccines *is* a beautiful thing. Why? They would say, "for the greater good." While I don't believe in evolution, most of the world does. I understand that vaccines are affecting our DNA, thereby weakening us. Could the big picture behind vaccines be some sick way to speed up the survival of the fittest that population control billionaires want so badly?

"Any upside to this dismal landscape? Well, the human species is subject to the same laws of natural selection as all other life on earth. Those with the weakest immune systems drop out. The immuno-suppressive effect of the immense viral load being forced onto the young through vaccines over long periods of time will certainly cause their line to have the least chance of replication of a viable progeny. It will dilute their DNA forever."[34]

"Mark Lappe talks about natural selection, survival of the fittest, and how vaccines have produced an artificial detour in our ability to evolve as a race, to refine our DNA. If that's true, the long-term survivors will be

the unvaccinated

those who are immune to vaccines."[35]

The racist theory of evolution, which was such an inspiration to Hitler and the genocidal like, appears to possibly be the motivation behind those who promote vaccines—population control under the guise of health. Most of us are consumed with making a living, paying off debt, raising our kids, trying to get healthy, and so on. A billionaire doesn't have debt to pay off, and they can pay people to raise their kids, pay personal chefs, etc. With billions of dollars and nothing to work toward, provoking the next stage of "human evolution" by devouring "*the poor secretly*" (Hab 3:14) could just be what warms the cockles of a cold, dead, population-control zealot's Mephistophelian heart.

"To do evil a human being must first of all believe that what he's doing is good… Ideology—that is what gives devildoing its long-sought justification and gives the evildoer the necessary steadfastness and determination. That is the social theory which helps to make his acts seem good instead of bad in his own and others' eyes, so that he won't fear reproaches and curse but will receive praise and honors."[36]

Ecclesiastes 8:12 **Though a sinner do evil an hundred times, and his days be prolonged, yet surely I know that it shall be well with them that fear God, which fear before him:** 13 **But it shall not be well with the wicked, neither shall he prolong his days, which are as a shadow; because he feareth not before God.**

(Read Job chapter 20 to see a wicked man's heritage from God.)

Hopefully, this situation with the COVID-19 vaccines and the discussions on ID chip implants have gotten many to realize that we are seeing the beginnings of Biblical prophecies start to unfold. This situation very well could be a forerunner to the mark of the beast, with vaccines being the vehicle behind its mass implementation. Interestingly, this has been predicted since smallpox vaccination.

"Brooklyn, New York, May 7, 1883.

Dear Sir:—I found the vaccination tyranny much more than sentiment on board the Adriatic. Aboardship, as everywhere, it has attained terrible proportions, which makes it probable that in the near future it will become the Great Terror that shall 'cause that as many as will not worship the image of the beast shall be killed,' and that 'no man may buy or sell save he that has the mark of the beast.'"[37]

To be clear: I'm not saying the COVID-19 vaccine is the mark of the beast. It appears that a future pestilence and its respective vaccine may be the way to carry out the mark on a worldwide scale. The mark could be showing that you have had a certain vaccine, that you are "safe." Most of us fell for the lies behind aborted fetal cell lines; it's a good thing true Christians will not be here for the mark because many of us would be fooled again.

There is a reason we have been programmed our whole lives to blindly accept vaccines, and that those who refuse them are idiots. (As far back as Job, it was said, "*the wicked…set me up for his mark*" in chapter 16 verses 11-12.) We've also been trained to disregard anything that someone labels a "conspiracy," even though the Bible uses that term multiple times. Not that every conspiracy is true, but they are not all automatically untrue, either.

"So what are Christians to make of all this conspiratorial kook-talk? First, they need to realize that, according to the Bible, 'conspiracies' are a NORMAL PART of a government operation when under the leadership of fallen man. … What this means, as unpalatable as it sounds in today's 'free-spirited' world, is that *there's a conspiracy afoot*. A conspiracy perpetrated by the 'rulers of the darkness of this world' (Eph 6:12) and joined in and carried out by secret counsels of men. The Bible shows that this is not only possibly, but *doctrinal*."[38]

Psalm 10:2 **The wicked in his pride doth persecute the poor…** 7 **His mouth is full of cursing and deceit and fraud: under his tongue is mischief and vanity.** 8 **He sitteth in the lurking places of the villages: in the secret places doth he murder the innocent: his eyes are privily set against the poor.** 9 **He lieth in wait secretly as a lion in his den: he lieth in wait to catch the poor: he doth catch the poor, when he draweth him into his net.** 10 **He croucheth, and humbleth himself, that the poor may fall by his strong ones.** 11 **He hath said in his heart, God hath forgotten: he hideth his face; he will never see it.**

Psalm 64:2 **Hide me from the secret counsel of the wicked; from the insurrection of the workers of iniquity:** … 5 **They encourage themselves in an evil matter: they commune of laying snares privily; they say, Who shall see them?**

(See also Job 24:14; Jer 5:26-29)

We must heed God's clear warnings of those who would *secretly murder the innocent*. It is not obvious why we have been trained to scoff at "conspiracy theories"?

The first two COVID-19 vaccines licensed (Moderna and Pfizer) are not made from an aborted fetal cell line, but they have been used to test these two vaccines. (According to information provided to Project Veritas, a whistleblower believes the Pfizer vaccine contains aborted fetal DNA, but the company is not giving full disclosure.[39])

"The mRNA COVID-19 vaccines produced by Pfizer and Moderna **do not require the use of any fetal cell cultures in order to manufacture (produce) the vaccine.** Early in the development of mRNA vaccine technology, fetal cells were used for "proof of concept" (to demonstrate how a cell could take up mRNA and produce the SARS-CoV-2 spike protein) or to characterize the SARS-CoV-2 spike protein."[40]

According to *Children of God for Life* (COG) there are seven COVID-19 vaccines listed as of March 2022: Moderna, Pfizer, AstraZeneca, Johnson & Johnson, Novavax, Corbevax, and Covaxin. As already shown, Moderna and Pfizer's vaccines are tested with aborted fetal cell lines. AstraZeneca and Johnson & Johnson actually propagate the virus in aborted fetal cells. Novavax, Corbevax and Covaxin are the newest vaccines on their website. A page titled "*Corbevax, Covaxin and Novavax*" has the most relevant information and is available at cogforlife.org. In summary, after reading the medical literature, COG affirms the use of aborted fetal cell lines in the development and testing of Corbevax and Novavax. After researching, they found no connection between Covaxin and abortion, but this vaccine has not been approved for public use. If any new COVID-19 vaccines are approved, visit *cogforlife.org* to see all vaccines made from aborted fetal cell lines. They frequently update their vaccine chart when new information becomes available.

"Are any COVID-19 vaccines available that are not tested or produced using a cell line derived from an aborted child? *No, for now, all vaccines for the COVID-19 virus being distributed are produced and/or tested with cell lines that originated from an aborted child.* Each person/family must make a prudential decision with an informed conscience and do what is best for you, your family, and your community. ... A COVID-19 vaccine that does not use aborted fetal cell lines in research or production is not expected in the foreseeable future."[41]

As of right now, many injuries and deaths have been reported and officially attributed to COVID-19 vaccines.

"The new CDC data through October 1, 2021, records 16,310 deaths besides 778,683 cases of injuries following COVID-19 shots. There are also another 23,712 permanent disabilities, 87,814 Urgent Care visits, 75,605 Hospitalizations, and 17,619 Life-Threatening injuries. That includes 2,102 fetal deaths from pregnant women who took a COVID shot."[42]

Because many have contracted this disease, and many are being coerced, or fooled, into receiving the COVID vaccines, here are some simple ways to protect from the damage caused by this illness and its vaccines. Black cumin seed and dandelion have both demonstrated to have protective properties.

Black Cumin Seed: "COVID-19, a global-pandemic binds human-lung-ACE2. ACE2 causes vasodilatation. ACE2 works in balance with ACE1. ... Nigellidine strongly binds to the spike-protein at the hinge-region/active-site-opening which may hamper proper-binding of nCoV2-ACE2 surface. ... The ACEs-blocking could restore Angiotensin-level, restrict vaso-turbulence in Covid patients and receptor-blocking might stop inflammatory/vascular impairment."[43]

"...black seed can "restrict vasoturbulence" in patients, while the receptor blocking could help to stop resulting inflammation and vascular impairment, especially helpful to the elderly and the co-morbid."[44]

Dandelion: "Infection of the lung cells using SARS-CoV-2 spike pseudotyped lentivirus particles was efficiently prevented by the extract and so was virus-triggered pro-inflammatory interleukin 6 secretion. ... SARS-CoV-2 is steadily mutating during continuous transmission among humans. This might eventually lead the virus into evading existing therapeutic and prophylactic approaches aimed at the viral spike. We found effective inhibition of protein-protein interaction between the human virus cell entry receptor ACE2 and SARS-CoV-2 spike, including five relevant mutations, by water-based common dandelion (Taraxacum officinale) extracts."[45]

Many believe spike protein transmission is occurring from the vaccinated to the unvaccinated due to the fact that women who have not received any COVID vaccinations, but have been in proximity to those who have, are reporting disturbances in their menstrual cycles.

Please note that black cumin is not recommended for use during pregnancy. But small amounts of dandelion are considered safe and are recommended to stop spike protein transmission.

"This means **infection of the lung cells using SARS-CoV-2 was efficiently prevented** by a water-based dandelion extract (basically a tea). It also means that dandelion tea used in moderation in pregnancy could **possibly prevent miscarriage** by blocking the spike protein from cross-reacting against [syncytin], which is vital in pregnancy."[46]

COVID's spike proteins are similar to syncytins, which is why drinking dandelion tea may be important for pregnant women, or those trying to conceive:

"Syncytins are physiologically expressed during pregnancy: they intervene in the development of the placenta, trophoblast differentiation, the implantation of the embryo in the mother's uterus and the immunosuppression of the mother's immune system to prevent allogeneic rejection of the embryo. Because of the similarity between syncytins and the spike protein of SARS-CoV-2, COVID-19 vaccine-induced antibody responses could trigger a cross-reaction against syncytins, causing allergic, cytotoxic and/or autoimmune side effects affecting human health and reproduction."[47] Roxana Bruno, Biochemistry PhD in Immunology

"Another good thing: a Journal of Functional Foods study shows **Common Dandelion root components safely** exhibit anti-oxidative and antiplatelet action. **Basically, this tells us that common dandelion herb** blocks **vascular damage** causing the strokes, heart attacks, migraines, blood clots in covid patients and covid vaccine recipients! With word in the news about thousands of women reporting miscarriages, disrupted menstrual cycles, and abnormal bleeding, this is good news."[48]

(Here is the study mentioned in the previous quote.[49])

Aside from the increase in miscarriages, there have also been increases in cancer rates and neurological disorders corresponding to the public's use of COVID-19 vaccines:

"On January 24, attorney Thomas Renz, a member of the America's Frontline Doctors (AFLDS) legal team, revealed to a panel that Wuhan coronavirus (Covid-19) 'vaccines' are extremely dangerous, despite constant reassurances from the government that they are 'safe and effective.' Three military doctors from the Department of Defense (DoD) who have access to vaccination data that has been withheld from the general public procured the information. They are Lt. Col. Theresa Long, Dr. Samuel Sigoloff and Lt. Colonel Peter Chambers. 'All three have given me this data in declarations that stated this is under penalty of perjury, we intend to submit this to the courts,' Renz said. What these three whistleblowers showed with the data is that miscarriages have increased by 300 percent over the past year, as have cancers. Neurological problems increased 1,000 percent during the same timeframe."[50]

I would like to mention that the Regeneron treatment recommended to COVID sufferers is also a product that uses stem cells created through in-vitro fertilization in its testing.[51] The biological materials required for in vitro fertilization were donated by "consenting individuals": they are still conceiving human life and terminating it at their convenience.

The Psalms repeatedly mention men who conspire to kill the poor secretly. We are also told the wicked will bring evil devices to pass (Psa 37:7), but there is something wicked man has tried to accomplish that will never succeed:

> Psalm 21:11 **For they intended evil against thee: they imagined a mischievous device, which they are not able to perform.**

My personal belief is that population control is the end that will never be met mentioned in the last verse.

> "Considering all this, one can conclude through common sense that these 'vaccines' are better suited for a tool of mass genocide. When are we going to wake up?"[52]

Noah and his family entering through the door of the ark that was then sealed by God (Gen 7:16) before the earth was flooded is a picture of salvation through Jesus Christ, the sealing of the Holy Ghost, and the rapture of the church. Soon the Son of Man will return. The rapture will happen before His Second Advent, before God's wrath is poured out on earth. The church will meet the Lord *in the air* (1Th 4:17); Christ does not physically return to the earth at the rapture. Rapture is a man-made term, and is also called the day of Christ (2Th 2:1-2). Our gathering together unto Christ is coming, and it seems as though it is coming *soon*. While many Christians are willingly ignorant of many scriptural truths, such as the heavens being *"of old"* by the Word of God (2Pe 3:5), we are told to not be ignorant of this *one thing* a few verses later in 2 Peter 3:8: "*...that one day is with the Lord as a thousand years, and a thousand years as one day.*" If you put the 6,000 years of *human* history next to the six days of creation, we are coming upon the seventh "day." In Genesis, God rested on the seventh day; the earth will have a time of rest when Christ reigns on earth in the millennium—for a thousand years (Rev 20:6). This world is becoming more wicked by the minute, just like the days of Noah (Mat 24:37); only when the Prince of Peace (Isa 9:6) comes will we have peace. **Jesus**

is coming back, and He will rule with a scepter of righteousness (Heb 1:8). This world claims to want peace, but peace is a product of righteousness (Isa 32:17), which this world does not want.

If you are not absolutely sure you are saved, repent and believe the Gospel: Jesus is the only way to be saved. The longsuffering of God waited in the days of Noah (1Pe 3:20), but He did what He had promised—destroyed the earth. God will do the same again, with fire rather than water; His longsuffering has a limit. **Now** is the accepted time, **now** is the day of salvation (2Co 6:2); there might not be a time to "get right with God" later. Repent and believe now, at the *accepted* time.

Chapter 12

Speak the Truth in Love

Ephesians 4:14 **That we henceforth be no more children, tossed to and fro, and carried about with every wind of doctrine, by the sleight of men, and cunning craftiness, whereby they lie in wait to deceive;** 15 **But speaking the truth in love, may grow up into him in all things, which is the head, even Christ:**

1 John 3:18 **My little children, let us not love in word, neither in tongue; but <u>in deed and in truth</u>.**

All parents want what is best for their children, and we have all certainly made mistakes. If you teach your children about Jesus and shepherd their hearts to come to a saving knowledge of Him, you have done what is most important in your child's life, and their eternity. If you are a parent, you have most likely let your child receive vaccines, and reading this information has potentially pricked your conscience. It is not your fault that someone lied to you, and you made a decision based on that lie. God knows who would have made the right choice had they been given the truth. God ponders the hearts, as it says in this interesting passage of Scripture:

Proverbs 24:12 **If thou sayest, Behold, we knew it not; doth not he that pondereth the heart consider it? and he that keepeth thy soul, doth not he know it? and shall not he render to every man according to his works?**

Also, if your children are saved, *they will get new bodies* (1Co 15:50-57). Any negative effects caused by vaccination will only last until the end of this life, which is only a vapor (Jas 4:14); a blessed hope, indeed (Tit 2:13).

Beware, fellow Christian, if this information has brought you to a crossroads, it is imperative that you choose wisely. When Jonah didn't obey God, he could not move forward with his spiritual life until he corrected his previous disobedience. Choosing to reject God's clear commandments will not bode well for your spiritual growth.

Psalm 66:18 **If I regard iniquity in my heart, the Lord will not hear me:**

Proverbs 28:9 **He that turneth away his ear from hearing the law, even his prayer shall be abomination.**

(See also Isa 1:15; Jas 4:17)

If you are a Christian who would consider yourself pro-life, but believe aborted fetal cell lines have been used for good, can you truly claim this title? You believe the exact same thing as those who identify as pro-choice: under certain circumstances, abortion is acceptable. Your standards might be much higher, but your position is not much different. One day, we will *all* stand before Christ (Rom 14:10), and while you might be able to fool yourself, but you cannot fool Him. Judge yourself and repent now (1Co 11:31), rather than when you're looking into Christ's eyes, which are as a flame of fire (Rev 1:14). Brothers and sisters, do not be fooled; "*your sin will find you out*" (Num 32:23).

Job 13:9 **Is it good that he should search you out? or as one man mocketh another, do ye so mock him?** 10 **He will surely reprove you, if ye do secretly accept persons.** 11 **Shall not his excellency make you afraid? and his dread fall upon you?**

All Christians have a choice between obeying or disobeying God's Word. Be warned, Christ told us that those who would heed His words would be given more; those who will not hear, what little they do have *will be taken away*.

Mark 4:22 **For there is nothing hid, which shall not be manifested; neither was any thing kept secret, but that it should come abroad.** 23 **If any man have ears to hear, let him hear.** 24 **And he said unto them, Take heed what ye hear: with what measure ye mete, it shall be measured to you: and unto you that hear shall more be given.** 25 **For he that hath, to him shall be given: and he that hath not, from him shall be taken even that which he hath.**

Is Jesus speaking about faith, or perhaps understanding of God's Word? Whatever Jesus is giving for hearing and obeying, we don't want to go without it.

Many Christians understand that the theory of evolution is not based on observable facts, yet is presented as though it is an undisputed truth. This is completely acknowledged, but when the same methods are used to push vaccines made from aborted fetal cell lines, many of those same God-fearing people fully support and believe these lies without any investigation. Anyone who puts pride aside and earnestly searches will see the truth. Upon researching, you will certainly relate as I did to an utterance from nearly 150 years ago: "Vainly do I seek to discover the advantages of vaccination."[1] Unfortunately, these words that are older than any man alive are also still true: "Indeed, I consider we are now living in the Jennerian epoch for the slaughter of the *innocent,* and the unthinking portion of the population."[2] Vaccines do not impart true immunity, man simply cannot do in a lab what God does in nature. Man can make some sort of "Frankenstein's monster" form of immunity; we break laws of nature to forcibly make a faulty, short-lived imitation of immunity that is accompanied by dire consequences.

"It (vaccination) endangers life, and in other cases leaves permanent impairment of health."[3]

I believe Christians should strive to absolutely avoid any kind of product made or tested with an aborted fetal cell line. Ultimately, what you will do is your decision, and your own business. While receiving morally-made vaccines is also your choice, it is a decision that ignores much of God's revealed wisdom. From injecting animal DNA to a "controlled" spread of sickness rather than *separation from* sickness, vaccination goes against God's wisdom in many ways. Personally, I will not let myself or anyone in my family receive any vaccine ever again. That is not to say that I think everyone should do what I'm doing, but everyone should be fully informed about what they are choosing.

Romans 14:5 **One man esteemeth one day above another: another esteemeth every day alike. Let every man be fully persuaded in his own mind.**

This verse is dealing with preferences, not on choosing to follow or ignore a clear command from God. While I believe it's clear Christians should diligently avoid aborted fetal products, what you do is between you and God. Furthermore, I'm not going to judge someone who received a vaccination to keep their job or military position to support their families. Many missionaries are required to take the COVID shot to continue their work. If I was called to be a missionary, I would do the same if it was absolutely necessary.

Man cannot fully understand the immune system or the viruses and animal DNA on which they are experimenting. Although otherwise claimed, man is "...bereft of complete and direct proof of how vaccines affect the human body."[4]

"Considered as a technical process, such a method (vaccination) is highly objectionable because it involves so many uncertainties. The living organism is an infinitely complex thing, when we consider the almost innumerable mechanical, chemical and vital processes going on within its constantly changing fluids and solids. There are no means of accurately registering and measuring all these activities; no means of determining exactly what these changes are; nor how they are modified by the introduction of the foreign morbid substance used."[5]

If the experts don't really know as much as they claim, and those who trust the experts do not investigate or think for themselves, how can this situation be anything other than the blind leading the blind (Mat 15:14)?

"There is no doctor or test that can predict whether or not vaccination will bring death, disability or the potentialities of future disease. Is the risk worth the dubious promise of immunity?"[6]

As Bible-believing Christians, there should be absolutely no fear in abstaining from vaccines made with aborted fetal cell lines (see 1Sa 12:24; 1Jo 5:3). The thought of declining vaccinations for my child once made me fearful. I now recognize that fear was, as one author said, "...little better than superstition,"[7] as is fear that unvaccinated children can harm vaccinated children by their mere presence. An unvaccinated child cannot spread diseases to which they've never been exposed.

"...it is the vaccinated children in this country whose immune systems are being systematically suppressed, whose inner milieu is made a viable medium for opportunistic organisms, now receiving more vaccines than any group of children in the history of the world. And who are thereby a true reservoir for disease and a threat to the unvaccinated."[8]

While some treat unvaccinated children as lepers, it is the vaccinated who are more likely to be sick, and to spread illness. I do not consider these children a threat to my child, nor do I fear them. If I'm truly trusting in God, what others choose to do should not frighten me. God has not given us a spirit of fear, but of power, love, and a sound mind (2Ti 1:7).

It may take "...an estimated average of seventeen years before evidence from scientific research is incorporated into day-to-day clinical practice,"[9] which explains why some in the medical profession are ignorant of some of these newer data on the dangers of vaccination. However, men have been speaking out against vaccination for *two hundred years.*

"I am convinced that vaccination is the greatest mistake and delusion in the science of medicine; a fanciful illusion in the mind of the discoverer; a phenomenal apparition, devoid of scientific foundation, and wanting in all the conditions of scientific possibility."[10]
—Dr. Joseph Herman, Principal Physician at the Imperial Hospital, Vienna, from 1858 to 1864

"Every intelligent person who takes the time to investigate this subject, will find abundant evidence in the published writings and public records of the advocates of vaccination, to prove its utter worthlessness, without reading a line of anti-vaccination literature. And if we could add to this all the suppressed facts, we would have a mass of evidence before which no vaccinator would dare to hold up his head."[11]
—Dr. Robert A. Gunn, MD, 1882

"It is the most outrageous insult that can be offered to any pure-minded man or woman. It is the boldest and most impious attempt to mar the works of God that has been attempted for ages. This stupid blunder of doctor-craft has wrought all the evil that it ought, and it is time that free American citizens arise in their might and blot out the whole blood poisoning business."[12]
—Dr. J.M. Peebles, MD, MA, PhD, 1900

"...both the alleged *protection* from attacks of disease and *mitigation* of its severity, when it does attack, are shown not only to be absolutely untrue, but to apply, in this case, to the absence of vaccination. ... I am firmly convinced that Edward Jenner saddled a legacy of disease and death upon the human race, and incidentally made $150,000 by the transaction; That the practice of vaccination has been the means of disseminating some of the most fatal and loathsome diseases, such as leprosy, cancer, syphilis, tetanus and tuberculosis; That vaccination is not only useless, but positively injurious..."[13]
—J. W. Hodge, M.D., 1902

"From its inception until the present day, the vaccination scheme has been an endless record of lies, deception, fraud, juggling statistics, and falsifying death certificates in order to preserve vaccination from reproach and to secure its continuation...and all this after more than a century of terrible experience, which has demonstrated that vaccination has killed more than smallpox, besides crippling and disfiguring millions more."[14]

"...our Creator, when He fashioned the human body, closed the blood stream from outside contamination. All objects entering the circulation must pass through intricate, but efficient, natural filters which bar out impurities. Yet man, with his hypodermic gadget, circumvents these filters, regardless of the Almighty's forethought. If the Creator had believed that those filters were not important, a valve leading into the blood circulation would have been provided with the same ease with which He has fashioned a human pump for the blood that will outwear, outbeat and outlast anything that man ever has, or ever will, conceive. ...a nutritional unbalance cannot be corrected by cow-pus or monkey-juice [smallpox and polio vaccinations]. All ailments result from such unbalance. No vaccine can supply what is missing or absorb poisons that cause the damage."[15]

"Vaccination is a grotesque superstition."[16]
—Dr. Charles Creighton, MD, MA

"My honest opinion is that **vaccination is the cause of more disease and suffering than anything I could name.**"[17]
—Dr. Harry R. Baybee

"The further I looked into it, the more shocked I became. I found that the whole vaccine business was indeed a gigantic hoax. Most doctors are convinced that they are useful, but if you look at the proper statistics and study the instance of these diseases, you will realise that this is not so."[18]

—Dr. Archie Kalokerinos, MD, PhD, AMM, MBBS, FAPM,

"Much of what you have been led to believe about immunizations simply isn't true. I not only have grave misgivings about them; if I were to follow my deep conviction in writing this chapter, I would urge you to reject all inoculations for your child."[19]
—Robert Mendelsohn, M.D.

"Often, when one objectively searches for information, facts are uncovered that can be in shocking contrast to the original understanding. The truth may be uncomfortable, inconvenient, and unpopular, but in the end, if it is the truth, it must be embraced regardless of the cost."[20]

If vaccines were as effective as is claimed, they would still not be worth their risks, given that much safer methods are available to keep our children healthy. As it happens, vaccines do not even have a record of efficacy to attempt to justify their continued injection into our children. This has always been true of vaccination.

"Even if there was any evidence to prove that vaccination was a prophylactic against small-pox, the appalling evils that have been and are still produced by it are sufficient to condemn the practice as a crime."[21]

No matter how appealing this world makes vaccination appear, the truth cannot be concealed from anyone who truly searches. Those who know the truth of vaccination will never again submit to this farce.

"...no man, with any self-respect, who has once seen through the stupid superstition, the shameless deceit of vaccination, will, without resisting to the uttermost, ever consent to the degradation of allowing the near and dear to him to be subjected to it, or lend a hand to the coercion of others."[22]

Remember these three main absolutely, undeniably true facts:

1. Vaccines did not cause any infectious diseases to disappear.
2. Some children will be maimed or killed by vaccines.
3. A child can still get diseases they've been immunized against.

These simple facts alone should keep parents from blindly following the recommended vaccine schedule.

Since I first decided to deviate from the government's vaccine recommendations with my child, I have not once read a single piece of factual information that has made me regret this decision. There have been some questions, and many things the media presented out of context that I couldn't answer until I did more research. The answers were always reassuring. It has been work, but it has been worth it.

"It takes months of agonizing research in most cases to de-program oneself from a lifetime of conditioning, and then a modest amount of moral fibre to actually take the initiative to sign the exemption form, thus freeing up the child's formative immune system from being an experimental petri dish for whatever neurotoxins the FDA decides to mandate this year."[23]

"As more and more parents are distrusting the mass conditioning from most sources, and actually informing themselves, the percentage of unvaccinated children in the US is increasing very slightly every year, despite

incessant media endorsement

school requirements

the course of least resistance

criminalizing the unvaccinated

The increase is astounding when one considers that to resist vaccination requires self-education - enormous personal effort, following up the spark of a new perception that hey, there might really be something to all this noise about vaccines, and I as a parent will do whatever is necessary to protect my child. Even if it means actually reading something."[24]

I have summed up my agonizing months of research for you. I have read accounts of children suffering and dying because of vaccines that they did not need, or could have been given later with less risk to the child. In the name of finding what is best for my child I have seen things and read things that I can't wait to forget. Many children and their families have suffered terribly in the name of "the greater good." I have taken much time to compile this information to warn others; I pray it will not be in vain.

Proverbs 24:10 **If thou faint in the day of adversity, thy strength is small.** 11 **If thou forbear to deliver them that are drawn unto death, and those that are ready to be slain;**

The vaccine industry was built on the bodies of murdered infants. That vaccination has saved *even one* life is highly debatable; that it has killed and maimed thousands upon thousands of innocents is an undeniable fact. It is a practice born of superstition and ignorance; it is forced upon mankind because of greed. Death is required in the manufacture of vaccines, and they continue to beget nothing less. "The Vaccination Monster"[25] should have been long dead but is instead alive and well, ***while many of our children are not.*** To sum the entirety of the vaccine issue: men are trying to play God and are failing miserably.

"We have indeed embraced the monstrous over the beautiful and good."[26]

We choose the monstrous ways of man over the beautiful and good ways of God, and our children are paying the price. The practice has proven to not be nearly as "safe" nor as "effective" as we've been told our whole lives, and upon studying the history of vaccination, it becomes clear that it has never been.

"The vaccination practice, pushed to the front on all occasions by the medical profession, and through political connivance made compulsory by the state, has not only become the chief menace and gravest danger to the health of the rising generation, but likewise the crowning outrage upon the personal liberty of the American citizen. ... Compulsory vaccination, poisoning the crimson currents of the human system with brute-extracted lymph under the strange infatuation that it would prevent smallpox, was one of the darkest blots that disfigured the last century."[27]

"Those whose interest it is to keep up the belief in the prophylactic value of vaccination against small-pox, have endeavored to make it appear that anti-vaccinators were bigoted and ignorant, and therefore not entitled to credence. The truth is that they are neither bigoted nor ignorant, but have been compelled to abandon vaccination, after giving the entire subject a careful and impartial investigation. ... With the conviction that they are right, and with the cooperation of many of the ablest minds the world has ever known, they can afford to smile at the invectives and ridicule of the interested, indifferent, and ignorant physicians, who find it necessary to do everything in their power to perpetuate the popular faith in vaccination."[28]

Many things could be said about those who choose not to vaccinate, most are probably untrue, but the biggest lie is that they are ignorant or unscientific. Anyone who has actually taken the time to inform themselves cannot call this movement one of ignorance.

"These comparisons emphasise that, regardless of how the medical establishment feels about anti-vaccinationists, it is important to understand that they have deeply held beliefs, often of a spiritual or philosophical nature, and these beliefs have remained remarkably constant over the better part of two centuries. The movement encompasses a wide range of individuals, from a few who express conspiracy theories, to educated, well informed consumers of health care, who often have a complex rationale for their beliefs, related to a 'mixture of world views held about the environment, healing, holism...and a critical reading of the scientific and alternative literature.'"[29]

This has always been a movement of truth and knowledge, propelled by those who believed in doing what is right and wanted to warn the public about its dangers. Remember, as one author accurately stated: truth can afford to wait the ages, but error would expire in the way.

Acts 20:27 **For I have not shunned to declare unto you all the counsel of God.**

Galatians 4:16 **Am I therefore become your enemy, because I tell you the truth?**

"It seems as if a man who could tolerate vaccination after these examples, would fiddle if the world was on fire, at least till the flames came around his own seat."[30]

Part IV

Ecclesiastes 1:18 **For in much wisdom is much grief: and he that increaseth knowledge increaseth sorrow.**

1 Corinthians 13:12 **For now we see through a glass, darkly; but then face to face: now I know in part; but then shall I know even as also I am known.**

"Vaccination is a monstrosity, a misbegotten offspring of error and ignorance; and, being such, it should have no place in either hygiene or medicine... Believe not in vaccination, it is a worldwide delusion, an unscientific practice, a fatal superstition with consequences measured today by tears and sorrow without end."

—Dr. Carlo Ruta, Professor of Materia Medica at the University of Perugia, Italy, 1896

Chapter 13

Give Place

Matthew 2:18 **In Rama was there a voice heard, lamentation, and weeping, and great mourning, Rachel weeping for her children, and would not be comforted, because they are not.**

"The child that was is no longer."[1]

Many claim that a parent reporting a vaccine reaction in their child is anecdotal evidence and therefore isn't "scientific." Any parent understands that they know their children better than anyone. It's very heartbreaking to read accounts of children developing normally, and then they stop. Their personalities completely change, and a parent watches their child turn into a different person. To tell these parents what they experienced isn't relevant evidence is shameful.

"When DPT [a group called Dissatisfied Parents Together] presents case histories, it is called anecdotal data, but when doctors do it, it is called clinical observation."[2]

"Mothers, who know and observe their children far more carefully than any doctor ever will, have reported for decades that their healthy babies become sick and sometimes die shortly after the doctor has given them a vaccination. But doctors have rejected, and continue to reject, their conclusions. ... For two thousand years, clinical observation and description of disease has been the raw material of medical science. But this raw material is apparently not to the taste of vaccine policymakers who dismiss as 'anecdotal' the stories of mothers and fathers who come forward with their vaccine-injured children."[3]

Please read one example of the many accounts we are told should be discounted because they are "anecdotal." It is heartbreaking and sobering to read, but I will tell you how to deal with the sorrow and anger inspired by these tragedies Biblically. The Bible tells us that sorrow is better than laughter: "*...for by the sadness of the countenance the heart is made better*" (Ecc 7:3). I once stopped researching vaccines because of the suffering I was seeing—but now, I do what God has told us, and it greatly helps me to deal with this present evil world (Gal 1:4).

Romans 12:19 **Dearly beloved, avenge not yourselves, but rather give place unto wrath: for it is written, Vengeance is mine; I will repay, saith the Lord.**

God has said He will repay, but Scripture has more to tell us on this subject. First, please read this story. This one example represents thousands upon thousands of similar stories of infant death, and reason with yourself as to why scientists discredit any negative publicity toward vaccines by discounting these experiences as irrelevant.

After Janet and Anthony's son Richie received his DPT shot, he became ill. Richie's older brother had already had a reaction, but Richie's pediatrician said his brother's reaction was "normal" and wanted to go ahead with his shots. Richie also had a similar reaction, the injection site became red, swollen, and he had diarrhea and vomited. The next day he cried, screamed, slept, and still had diarrhea. That night he became cold, his breaths were accompanied by an odd sighing sound, and his pupils stopped adjusting to light. Janet called the pediatrician who did not seem overly concerned but said his parents could take Richie to the hospital if they wanted to. As they were trying to get their two children to the car to go to the hospital, Richie stopped breathing. Janet tried to resuscitate her baby who vomited, and died shortly afterward.

"'The ambulance came and the paramedics continued CPR. I knew my baby was gone, I kept screaming he was dead. When he had stopped breathing in my arms, all of the sudden, the whole day came together like one big nightmare. He had been dying all day, and I didn't even know it.' Richie had died thirty-three hours after he received his first DPT shot. … 'I thought to myself…of course I would have called the doctor in a minute if Richie had not had the DPT shot the day before and I had not assumed he was having a normal reaction to the shot. We have been told that these reactions to vaccines are normal. We are so conditioned to that idea and to the idea that our doctor's word is to be trusted without question that we don't think for ourselves. I am a nurse. I watched my son die that day, and I didn't even know what was happening until it was all over.'"[4]

Janet was given some papers on SIDS, but vehemently opposed that her son died of SIDS, and was adamant that he died because of the DPT shot he had just received. When she asked why the coroner would not put on Richie's death certificate that a vaccine had caused her son's death, "He said he couldn't write down on the death certificate that Richie had died from a DPT reaction because 'the state's standing on immunizations would be in an uproar.'"[5]

"A part of me died with Richie, I can't bring him back. I can only continue my life. These doctors and officials in the government, who keep talking about the benefits and risks of this vaccine, better take fair warning. My baby may be just another statistic to them, but he was my child and there is nothing more powerful than a mother's fight for her child. There is no war stronger than a mother's fight for her child. I will fight no matter what I have to do and no matter how long it takes to keep this from happening to other babies. I will not let the cause of my baby's death be forgotten.' Janet looks away. Her voice is angry, but her eyes are wet with tears. Carefully, almost lovingly, she takes Richie's autopsy report out of an envelope. 'This is all I have left of him,' she says. 'This report and the pictures we took the day he was christened.' She reads the autopsy report, as if reading it one more time will lead her from that dark land where a mother grieves for the child she will never hold again."[6]

Another mother articulated a sad truth:

"…God gave me a perfect child. I was so happy when he was born. He was so beautiful, with ten toes and ten fingers. God gave me a perfect child and man, with his own ways, damaged God's perfect work."[7]

We have all submitted to vaccination, with a vague, nebulous understanding that they can be harmful to some, but of course, we think nothing bad will happen to us, nor to our children.

"Without the truly informed consent of the American public, these vaccine policymakers determined that it was necessary and morally acceptable to sacrifice an unknown number of children for the theoretical well-being of the rest of the society. And then vaccine-damaged children and their parents were left to fend for themselves. Perhaps it was easy to do that when one assumed somebody else's child would be sacrificed. Again and again, mothers of vaccine-damaged children have said, 'When it happens to your child, the risks are 100 percent."[8]

Man is slow to learn that his methods to prevent disease are worse than what is supposedly prevented.

"The number of well children who have suffered mental or physical disturbance as a result of vaccination may have to be counted not in hundreds, as in the United Kingdom, or in thousands, as in the rest of the Western world, but in hundreds of thousands all over the world. This is a heavy price to pay for marginal and temporary protection against a disease that seldom nowadays threatens the life or health of well children."[9]

Another mother, whose child was handicapped because of a vaccine reaction, wisely said:

"I learned a long time ago that I would destroy myself if I let anger and fear take over my life. I'm trying to channel it. I want to try to make the lives of handicapped children and their parents more bearable. I want to get information out to parents about the vaccine [DPT] to lessen the chances of other kids being hurt like Sherry was hurt."[10]

Psalm 56:8 **Thou tellest my wanderings: put thou my tears into thy bottle: are they not in thy book?**

If you are the parent of a child who died because of vaccines, please come to a saving knowledge of Jesus Christ, and see your child again. These children are with Him, and those who have wronged you will have to answer to the just Judge of the universe. God has seen every secret thing (Ecc 12:14), and *has counted your tears*. Precious in the sight of the Lord is the death of His saints; the deaths of little ones (Luk 17:2) are certainly regarded by the Most High (Ecc 5:8), and they will not go unanswered.

Psalm 116:9 **I will walk before the LORD in the land of the living.** 10 **I believed, therefore have I spoken: I was greatly afflicted:** 11 **I said in my haste, All men are liars.** 12 **What shall I render unto the LORD for all his benefits toward me?** 13 **I will take the cup of salvation, and call upon the name of the LORD.** 14 **I will pay my vows unto the LORD now in the presence of all his people.** 15 **Precious in the sight of the LORD is the death of his saints.**

It is claimed that because there is suffering, there is no God; or that He's not good if He does exist. Many mistakenly blame God for their pain in life—pain that is a consequence of rejecting God and His words. It is not God's will that children be afflicted nor grieved (Lam 3:33). In the book of Job, the things that happened to him did not happen because God was doing them. God did lift His protection over Job, but the things that happened were because of Satan. We live in a world that Satan is the god of (2Co 4:4); these things happen because of him, and our own sin, not God. One day we will see the end results of God's amazing grace, mercy, goodness, and justice. While doing evil is never allowed in Scripture, God has used some of these things for good—to them that love God and are called according to His purpose (Rom 8:28): children are His; they are an heritage *of the LORD* (Psa 127:3). Many children would not have gone to heaven had they lived their whole lives and chosen to reject Christ. God said He would forget the children of those who forgot His laws (Hos 4:6). If a people has forgotten God's law, how likely is it that these children would have come to a saving knowledge of Him? The Bible makes it very clear: the death of a child is a terrible thing. God is good, and will not allow this to go unanswered. God is not responsible for the wickedness of this world, He has given everyone the freedom to choose what they will do, and that is why terrible things happen. This world is wicked because of man, but God will set everything right. God's understanding is higher than ours—trust Him. This life is short (Jas 4:14), and eternity is **forever**.

Isaiah 45:22 **Look unto me, and be ye saved, all the ends of the earth: for I am God, and there is none else.**

Isaiah 35:4 **Say to them that are of a fearful heart, Be strong, fear not: behold, your God will come with vengeance, even God with a recompence; he will come and save you.**

Are the boys and girls playing in the streets of Christ's Kingdom (Zec 8:5) the children that did not get to live their whole lives throughout history? Read this verse from Deuteronomy, and observe the use of the words *"that day,"* which references the Day of the Lord; also notice the innocence of the children mentioned:

Deuteronomy 1:39 **Moreover your little ones, which ye said should be a prey, and your children, which in that day had no knowledge**

between good and evil, they shall go in thither, and unto them will I give it, and they shall possess it.

I can't help but remark that the reason the Israelites disobeyed God's command was because they feared for their children's lives. (See also Num 14:31.) Do all of the children who died in innocence get to grow up in the Millennial Kingdom, a perfect place that Jesus rules? Perhaps children originally born into a world where their own mothers consented to their deaths will instead live in a world where even animals will not harm them (Isa 11:8)? I eagerly await the day I meet these children whose sufferings I have read, and see them laughing and playing in a world ruled by the One who wanted the children to come to Him (Mar 10:14).

Psalm 90:3 **Thou turnest man to destruction; and sayest, Return, ye children of men.**

Jeremiah 30:19 **And out of them shall proceed thanksgiving and the voice of them that make merry: and I will multiply them, and they shall not be few; I will also glorify them, and they shall not be small.** 20 **Their children also shall be as aforetime, and their congregation shall be established before me, and I will punish all that oppress them.**

Isaiah 54:13 **And all thy children shall be taught of the LORD; and great shall be the peace of thy children.** 14 **In righteousness shalt thou be established: thou shalt be far from oppression; for thou shalt not fear: and from terror; for it shall not come near thee.**

In the gospels, Jesus has a habit of healing children and giving them back to their parents: Luke 7:15, 8:56, and 9:42. Will we see this happen again?

When our obedience is fulfilled, God will be ready to avenge all disobedience (2Co 10:6). Proverbs 22:22-23 says the Lord will plead the cause of the poor, and spoil those that spoiled them. The heathen will be caught in the net they made, and snared in the work of their own hands; the poor and needy will not always be forgotten; Christ will judge the world: "*that the nations may know themselves to be but men. Selah*" (Psa 9:15-20). God will render every man according to his deeds

(Rom 2:6). This is what will happen to those who devour the poor secretly:

Habakkuk 3:12 **Thou didst march through the land in indignation, thou didst thresh the heathen in anger.** 13 **Thou wentest forth for the salvation of thy people, even for salvation with thine anointed; thou woundedst the head out of the house of the wicked, by discovering the foundation unto the neck. Selah.** 14 **Thou didst strike through with his staves the head of his villages: they came out as a whirlwind to scatter me: their rejoicing was as to devour the poor secretly.**

We are told to give place unto wrath. The Bible also tells us that Christ's saints have the honor of participating in the repaying of vengeance.

Psalm 149:5 **Let the saints be joyful in glory: let them sing aloud upon their beds.** 6 **Let the high praises of God be in their mouth, and a twoedged sword in their hand;** 7 **To execute vengeance upon the heathen, and punishments upon the people;** 8 **To bind their kings with chains, and their nobles with fetters of iron;** 9 **To execute upon them the judgment written: this honour have all his saints. Praise ye the LORD.**

Through Christ's blood, born-again believers are made saints (Rev 1:5-6), and in the future we will rule and reign with Him (Rev 5:10). Give place to wrath (Rom 12:19), perhaps you'll do something with it in the future. (See Luk 19:11-27.)

It is better that all men repent and come to Christ, but the space to repent (Rev 2:21) that God gives men is running out. I've read of so much suffering, so much death, so much injustice, so many people being oppressed, and the poor being devoured secretly. These things that happened and resulted in children being harmed—there is no excuse; it is wicked. However, perhaps many will get to the Millennial Kingdom because they died in innocence (Psa 90:3). Those who have set up these world systems that kill and maim will answer for their crimes. According to Jesus, it is better for a man to die than to offend one of His children—because there are *eternal consequences* for such wickedness. Their eternity will be much worse than if they had not harmed children.

Mark 9:42 **And whosoever shall offend one of these little ones that believe in me, it is better for him that a millstone were hanged about his neck, and he were cast into the sea.**

After discovering how wicked this world really is, what I await with the most anticipation is to see all of the wickedness *undone* (Psa 112:10b; Pro 10:28). I believe every child slain in the name of "the greater good" will be made as though no evil had ever come to pass, and those who live to secretly kill the innocent and the poor will see all of their wicked "accomplishments" completely dissolved. Christ will make all things new, and these men will receive their just wages for their evil life's work that didn't amount to anything; *not even one* innocent child will perish (Mat 18:14; Luk 12:6) despite their snares and devices.

Revelation 21:5 **And he that sat upon the throne said, Behold, <u>I make all things new</u>. And he said unto me, Write: for these words are true and faithful.** 6 **And he said unto me, It is done. I am Alpha and Omega, the beginning and the end. I will give unto him that is athirst of the fountain of the water of life freely.**

This next verse was written to the Jews, as are most of the verses in this section about children returning and living in the Kingdom. I don't know when those who have died in innocence throughout time will be seen again. I do know the Judge of all the earth will do right (Gen 18:25); the Man who wanted the children to come to Him is going to make all things new (Rev 21:5); and Christ is going to reconcile all things in heaven and earth *to Himself* (Col 1:20).

Jeremiah 31:15 **Thus saith the LORD; A voice was heard in Ramah, lamentation, and bitter weeping; Rahel weeping for her children refused to be comforted for her children, because they were not.** 16 **Thus saith the LORD; Refrain thy voice from weeping, and thine eyes from tears: for thy work shall be rewarded, saith the LORD; and they shall come again from the land of the enemy.** 17 **And there is hope in thine end, saith the LORD, that thy children shall come again to their own border.**

Acts 26:8 **Why should it be thought a thing incredible with you, that God should raise the dead?**

Chapter 14

In Conclusion—Mercy and Truth

Jeremiah 17:14 **Heal me, O LORD, and I shall be healed; save me, and I shall be saved: for thou art my praise.**

Jeremiah 33:6 **Behold, I will bring it health and cure, and I will cure them, and will reveal unto them the abundance of peace and truth.**

The only Child God was willing to sacrifice was His Own. Through Christ's sacrifice, man became redeemable. If you are not saved, the only plague you need to worry about is death. The only cure you need is the cure for the eternal consequences of your sin. God gives length of days for ever and ever (Psa 21:4); in the way of righteousness there is no death (Pro 12:28). By *His* stripes we are *healed* (Isa 53:5).

It is claimed that the Bible is an unscientific book full of superstition and contradictions; this is simply untrue. If you take the time to examine the scientific truths of the Bible,[1] the prophecies in the Bible that have been historically recorded and were fulfilled exactly as the Scriptures predicted,[2] and the other historical accuracies in the Bible,[3] it becomes clear that it is a supernatural Book.[4] While it is certainly true that some translations *create contradictions* by changing God's words, in the faithful English translation of God's Word, the King James or Authorized Version, no legitimate contradiction has ever been identified.[5]

We briefly looked at some of the scientific truths of the Bible in Chapter 3; and throughout this book we have shown the accuracy of Scripture. These are only *some* of the truths contained in the Bible. There are hundreds of other prophetic and historical truths as well. It is mathematically impossible that man is the source of all this accurate information. Men may have penned the words, but God *inspired* the words (2Ti 3:16; 2Pe 1:21; Mar 12:36; Act 28:25). The Bible contains so much truth that it's impossible that it is not a supernatural book. It is God's Word, and it is true. What it tells us about death and hell are true. What it proclaims about a Savior is true. Jesus is the only way to eternal life (Jhn 6:68).

If you do not believe in God, I challenge you to investigate this belief the same way you have sought truth in other areas of life. As you read this, processing it with a mind science can't explain and a conscience that can't be quantified, do you have *faith* in the racist, sexist, unproven theory of evolution that has inspired several mass murderers? Are you trusting in something so illogical that it goes against scientific theories that *have been proven*? Do you not believe in God because there are "*...oppositions of science falsely so called*" (1Ti 6:20) as *He foretold* thousands of years ago? Will you not believe in God because it seems foolish, as *God chose it to be?*

1 Corinthians 1:18 **For the preaching of the cross is to them that perish foolishness; but unto us which are saved it is the power of God.** 19 **For it is written, I will destroy the wisdom of the wise, and will bring to nothing the understanding of the prudent.** 20 **Where is the wise? where is the scribe? where is the disputer of this world? hath not God made foolish the wisdom of this world?** 21 **For after that in the wisdom of God the world by wisdom knew not God, it pleased God by the foolishness of preaching to save them that believe.**

"In Isaiah 41:23 the prophet hurled out the challenge to heathen gods: 'Show the things that are to come hereafter, that we may know that ye are gods.' God has accepted this challenge. He has predicted multitudes of events to happen in the future. They have come true exactly as predicted, even though in some cases thousands of years were involved for the fulfillment. God has proved that He is our supernatural God with all wisdom. We have no alternative but to believe."[6]

If you already understand much of this information about vaccines but believe in evolution, please acknowledge the parallels. An unproven theory is promoted as fact; this theory goes against scientific laws, and it is deemed foolish not to believe in it. If you cannot believe parents will submit to letting their children receive vaccines without researching the matter, are you risking *your soul* in the same way? The theory of evolution claims that certain races are more favorable than others—it's in the title, you can look for yourself—and that women are less evolved than men. Please do not trust your eternity to an unproven theory that has never been observed. (Microevolution, the small changes within a species, has been observed; but macroevolution, changing from one species to another, has not.) Evolution and the second law of thermodynamics cannot both be considered scientific facts. The second law is observable and provable; the theory of evolution is atheistic fan fiction.

"*...evolution and the second law cannot both be true.* As for the second law of thermodynamics, it has been confirmed by numerous and varied scientific tests, while evolution is—and always has been—a theoretical model not even capable of scientific test. ...evolution is in trouble. No matter how enthusiastically it tries to sell its speculations, something in nature keeps standing up and saying No! The two great universal principles of thermodynamics—energy conservation and deterioration —bear witness to the scientific necessity of an original creation. ... In short, the biblical record, once dismissed in favor of evolution, is now seen to rest upon scientific foundation. ... If one must choose, it would seem more rational to go with science."[7]

Evolution is a belief that takes much more faith than Biblical Christianity.

"All of these are not more than fairy stories for adults, full of metaphysical self-defeating contradictions into which unbelieving scientists have fallen because they have rejected and/or ignored God's revelation of Himself. ... Thus both sides are going through life practicing their beliefs and in so doing are being 'religious.' One side honestly admits this; the other side is self-deceived and does not so concede. ...the fossil record which is offered by evolutionists as physical proof of their ***belief*** is totally against them. Not only are the transitional forms missing between all major animal and plant groups, the presence of which would be necessary to maintain such an hypothesis [of evolution-the author does not believe evolution merits theory status], the gaps are of an unimaginable magnitude. The fossil record is one of mass destruction, death, and burial (almost exclusively by water and its suspended sediments), not that of one kind transforming into another. Contrary and hostile to the evolutionary scenario, the fossil record not only reveals a systemic absence of transitional forms, it reveals the sudden explosive appearance [The Cambrian Explosion] of highly diverse forms of life without any preceding intermediate types. ... In short, the laws of thermodynamics tell us that evolution could not happen, and the fossil record tells us that evolution did not happen - and that is real science."[8]

The Cambrian Explosion mentioned above has never been explained, except by the Bible. Also, there is not just one "missing link": all of the "links" between every single form of life on earth are "missing."

"Physical degeneration, and not evolutionary improvement, is the story of life on this planet. ...where chromosome and gene mutations occur, they are almost always detrimental, rather than beneficial."[9]

2 Timothy 4:3 **For the time will come <u>when they will not endure sound doctrine</u>; but after their own lusts shall they heap to themselves teachers, having itching ears;** 4 **And they shall turn away their ears from the truth, and shall be turned <u>unto fables</u>.**

"Human intelligence cannot be the product of chance evolution. Man did appear suddenly—at the top, not at the bottom. Man is a created artifact, far more wonderful than any computer. He was carefully planned and endowed with the gift of language and the most amazing intellect, as well as a feeling for handcrafts and technology. And something else was implanted within man: both the capacity and the need to communicate with his Maker, the prime Intelligence. Consequently, there is a part of every person which is restless, seeking unattainable goals, yet experiencing futility and emptiness until it finds identity and peace with the Creator. And that makes all the difference."[10]

What follows is a borrowed illustration. Would you sell one of your eyes for a million dollars? What about both of them for a billion? Our eyes are precious to us, and we would not go without our sight for any amount of money. The eye is only the "window to the soul" (see Mat 6:22-23); how much more should you value your eternal soul? Please do not leave this matter uninvestigated. Look at the scientific truths of the Bible, the prophecies of the Bible that have been fulfilled, and the Bible's answers to life's biggest questions that evolution cannot answer. Simply looking at the history of the Jewish people and how it aligns with Biblical prophecy shows the divine inspiration of the Bible. There really is abundant *scientific* proof that attests to the truth of Scripture. Seek God, and do not go through life not knowing what comes after it. God is a rewarder of them that diligently seek Him, but *you must believe that He is* (Heb 11:6).

"More than any document, the Bible was assailed as a collection of fanciful myths. Yet, to the embarrassment of the critics, archaeological discoveries proved time and again that the fabled cities, mythical persons and impossible events were true and reliable reporting in every detail. Indeed, the Bible can now be regarded as the most accurate and trustworthy source of history we possess."[11]

1 Chronicles 28:9 **And thou, Solomon my son, know thou the God of thy father, and serve him with a perfect heart and with a willing mind: for the LORD searcheth all hearts, and understandeth all the imaginations of the thoughts: <u>if thou seek him</u>, he will be found of thee; but if thou forsake him, he will cast thee off <u>for ever</u>.**

1 John 5:20 **And we know that the Son of God is come, and hath given us an understanding, that we may know him that is true, and we are in him that is true, even in his Son Jesus Christ. This is the true God, and eternal life.**

Intellectually acknowledging there is a God will not take anyone to heaven; the Bible tells us the devils believe (Jas 2:19). Man must turn (Act 26:18) to God and trust in Christ's sacrifice (Act 20:21). If you are trusting in your good works (Isa 64:6) to save you, please understand that is an **insult to God**, *who gave up his Son* as the **only satisfactory payment** (propitiation) for our sins.

1 John 4:10 **Herein is love, not that we loved God, but that he loved us, and sent his Son to be the propitiation for our sins.**

To reject such a precious gift because we will not humbly accept that we are powerless to save ourselves is foolishness. It is folly to think one can earn their way to heaven when God has sacrificed His Son to accomplish this for us (Gal 2:21). How shall we spurn such a gift?

Hebrews 2:3 **<u>How shall we escape, if we neglect so great salvation</u>; which at the first began to be spoken by the Lord, and was confirmed unto us by them that heard him;**

Romans 10:3 **For they being ignorant of God's righteousness, and going about to establish <u>their own righteousness</u>, have not submitted themselves unto the righteousness of God.**

The only way to heaven is the way which God chose. Any man who rejects this gift (Eph 2:8-9), will be given exactly *what he has chosen*: an eternity without God, paying for his sins. God is not willing that any should perish, but that all should come to repentance (2Pe 3:9), but He will also not force any person to love Him.

John 8:47 **He that is of God heareth God's words: ye therefore hear them not, because ye are not of God.**

If you consider yourself a Christian, but do not remember a **point in time** when you repented and confessed to God that you trust in His Son Jesus as the only payment for your sins (Jhn 14:6), please do so (Phl 2:12).

Proverbs 16:6 **By mercy and truth iniquity is purged: and by the fear of the LORD men depart from evil.**

God's word tells us that iniquity is purged by mercy and truth. Man has nothing to do with the mercy part of this equation; man must submit to God's truth to purge iniquity from his life. Knowing that we will be completely forgiven helps us to turn to God (Rom 2:4). (If you're saved, you're already forgiven; asking God for forgiveness is for the benefit of *our* hearts and spirits.) If you have used any kind of aborted fetal medicine ignorantly, or if you were fooled to believe it was acceptable, turn from the evil ways of this world, and go to the one true God, in whom is no darkness (1Jo 1:5), variableness, nor shadow of turning (Jas 1:17). The wicked king Manasseh sacrificed his own children to false gods (2Ki 21:6; 2Ch 33:6). When he greatly humbled himself and sought the Lord, God answered his prayer (2Ch 33:12-13)! Whoso confesses and forsakes his sins shall have mercy (Pro 28:13). God will abundantly pardon those who forsake wickedness (Isa 55:7). We must come to God with our whole heart, and we must acknowledge our sin (Jer 3:10-13). Christ paid for every sin (1Jo 2:2), but those who did not want their sins covered (Rom 4:7) this way will get what they have chosen: the wrath of God (Rev 14:10).

Hebrews 10:31 **It is a fearful thing to fall into the hands of the living God.**

2 Thessalonians 1:7 **And to you who are troubled rest with us, when the Lord Jesus shall be revealed from heaven with his mighty angels,** 8 **In flaming fire taking vengeance on them that know not God, and that obey not the gospel of our Lord Jesus Christ:**

The only hope any sinner has is to repent and trust in Christ—repentance *is required*. Repentance is not a work, it is changing your mind from what you once believed to what God has declared. To repent is to turn from darkness to light; from Satan to God (Act 26:18). Whatever you are trusting in for your soul's salvation—good works, baptism, church membership, a weekly Eucharist, being a "good" person—let it go and cling to the cross (Col 2:14). The thief on the cross was sentenced to death for a life of wickedness; he had no time to do good works, join a church, or get baptized. All he did was believe in Jesus (Luk 23:39-43), and he clearly knew to whom he was speaking. When you ask Christ to forgive your sins, you must understand who He is. He is God, and there is no salvation in any other (Act 4:12). That is why the Bible tells us repentance is so important (Act 17:30): we must turn from trusting ourselves to trusting the One True God and what He has done to save us.

Luke 13:3 **I tell you, Nay: but, except ye repent, ye shall all likewise perish.**

It's important to note that no man will be condemned because of his wicked deeds; there will be condemnation because of what he did not do. All men are wicked, and all are capable of committing abominable acts. Christ will forgive any man who comes to Him in the way He has instructed man to approach Him (Jhn 3:15). Woe to those who will not accept this free gift of salvation. Woe to those to secretly devour the poor; woe to those women who have aborted their children; woe to the doctors who performed these operations. As evil as these actions are, God is willing to forgive them, but man must repent and believe that Jesus is Lord, and trust in His payment for their sins (Rom 10:9; 1Co 15:3-4): only His sacrifice is acceptable to God.

Fellow Christian, if you are truly saved, *nothing* is greater than Christ's love (Rom 8:38-39), promises (Gal 3:14), or the seal of the Holy

Spirit (Eph 1:13). Even if we turn from Christ after salvation, He must keep His word; He cannot deny Himself.

2 Timothy 2:13 **If we believe not, yet he abideth faithful: he cannot deny himself.**

If it were up to man to keep his salvation, it would surely be lost; Christ keeps our salvation, He confirms *us "unto the end"* (1Co 1:8). According to the Bible, if we offend in one point, we are guilty of the whole law (Jas 2:10); one sin after salvation would make us guilty of all. It is not Scripturally possible for man to keep his salvation. It is possible for men to forget they were purged from their sins (2Pe 1:9) and to live as though they are unsaved—but they are saved *"yet so as by fire"* (1Co 3:15). Truly born-again believers cannot have their salvation rescinded (Jhn 10:28), but we shall all stand before the judgment seat of Christ and *give an account* (Rom 14:10-12). After the new birth, the soul is not in danger by ignoring God's Word, but our rewards are (Rev 3:11).

2 Timothy 2:11 **It is a faithful saying: For if we be dead with him, we shall also live with him:** 12 **If we suffer, we shall also reign with him: if we deny him, he also will deny us:**

2 Timothy 2:19 **Nevertheless the foundation of God standeth sure, having this seal, The Lord knoweth them that are his. And, Let every one that nameth the name of Christ depart from iniquity.**

Psalm 73:24 **Thou shalt guide me with thy counsel, and afterward receive me to glory.**

Jesus is coming; time is short. Tell others the good news, and spread the gospel. We are watching the beginnings of the prophetic Book of Revelation come to pass. As our hope is not in this world, these things should not be overly troublesome to us. Tell others about the only way to have eternal life, through our Lord Jesus Christ. Read, study, learn, live, and love God's Word. God's words are righteous and completely pure (Psa 119:140). They are preserved by Him (Psa 12:7), esteemed above His very name (Psa 138:2), and more sure than the audible voice of God (2Pe 1:18-19). The Bible *is* God's inspired Holy Word (2Pe 1:21); it is true from the beginning (119:160) and is settled forever in heaven (Psa 119:89). Trust your life, soul, and children to its truth.

Ecclesiastes 12:13 **Let us hear the conclusion of the whole matter: Fear God, and keep his commandments: for this is the whole duty of man.**

Daniel 12:3 **And they that be wise shall shine as the brightness of the firmament; and they that turn many to righteousness as the stars for ever and ever.**

Additional Resources

Proverbs 19:20 **Hear counsel, and receive instruction, that thou mayest be wise in thy latter end.**

Note: Any book or documentary that tells the truth about vaccines will be falsely labelled as "misinformation," "unscientific," "quackery," and "anti-vaccine."

Complete Lists of all Aborted Fetal Vaccines, Products, and Ethical Alternatives:
https://cogforlife.org/wp-content/uploads/vaccineListOrigFormat.pdf
https://cogforlife.org/wp-content/uploads/fetalproductsall.pdf

Vaccine Information Websites:
National Vaccine Information Center, nvic.org
VIC Fellowship (Vaccine Information Coalition), vacinfo.org

Other recommended blogs/websites:
https://thetruthaboutvaccines.com
https://deeprootsathome.com

Recommended Books:

Miller's Review of Critical Vaccine Studies

Vaccine Epidemic: How Corporate Greed, Biased Science, and Coercive Government Threaten Our Human Rights, Our Health, and Our Children.

Dissolving Illusions: Disease, Vaccines and the Forgotten History

There is also a "History of Vaccination" series with nearly 30 books, they are available in digital format on Amazon and Google. They are all very inexpensive or free. Many of the older books sourced for this compilation are expensive to buy in print—this is a very affordable way to learn much about vaccination.

Vaccine information packets and compilations

1. Deep Roots at Home—Vital Vaccine information *For Moms & Dads To Make An Informed Choice* https://deeprootsathome.com/wp-content/uploads/2019/01/Deep-Roots-at-Home-E-Book-2019.pdf
2. Flu Vaccine for All: A Critical Look at the Evidence https://www.medscape.com/viewarticle/855937_1
3. The Denial of Adverse Event Risk Following Immunization and the Loss of Informed Consent - A Perspective https://www.actascientific.com/ASPE/pdf/ASPE-02-0045.pdf

Documentaries

The Truth about Vaccines Docu-Series hosted by Ty Bollinger. The first episode is available for free here: https://go2.thetruthaboutvaccines.com/docuseries/episode-1/

Vaccine Documentaries available for free on YouTube:

1. Vaccination – The Silent Epidemic
"Silent Epidemic; The Untold Story of Vaccines Movie"
https://www.youtube.com/watch?time_continue=3&v=K1m3TjokVU4&feature=emb_logo

2. The Greater Good
"The Greater Good - Real Stories of Vaccine Injury"
https://www.youtube.com/watch?v=eU32Ojf8-rA

3. Shots in the Dark
"Shots In The Dark Silence on Vaccine"

https://www.youtube.com/watch?v=fVwLo3lmKyo&feature=emb_lo go

4. Vaccination The Hidden Truth
"Vaccination: The Hidden Truth"
https://www.youtube.com/watch?time_continue=1&v=cqsT5EoIk8U &feature=emb_logo

5. Vaccine Nation
"Vaccine Nation (Documentary)"
https://www.youtube.com/watch?v=bLk641P8CE4

6. Lethal Injection
"Lethal Injection: The Story Of Vaccination"
https://www.youtube.com/watch?v=7hITYIT02rA&feature=emb_log o

Sources

Introduction

[1]Hodge, John W. *Vaccination Superstition: Prophylaxis to Be Realized through the Attainment of Health, Not by the Propagation of Disease.* Scholar Select, 2015. pp. 11-12.

[2]O'Shea, Tim. *Vaccination Is Not Immunization.* 4th ed., Immunition Ltd, 2015. p. 13.

[3]Peebles, J. M. *Vaccination, a Curse and a Menace to Personal Liberty: with Statistics Showing Its Dangers and Criminality.* Forgotten Books, 2018. pp. 5, 260.

Chapter 1 Thou Shalt Not Kill

[1]Joan C. Thicke, Darline Duncan, William Wood, A. E. Franklin and A. J. Rhodes. "Cultivation of Poliomyelitis Virus in Tissue Culture; Growth of the Lansing Strain in Human Embryonic Tissue," *Canadian Journal of Medical Science,* Vol. 30, pp. 231-245. Retrieved from Children of God for Life, "Forsaking God for the Sake of Science," https://cogforlife.org/2012/06/13/polioperversion/.

[2]Thomas H. Weller, John F. Enders, Frederick C. Robbins and Marguerite B. Stoddard. "Studies on the Cultivation of Poliomyelitis Viruses in Tissue Culture: I. The Propagation of Poliomyelitis Viruses in Suspended Cell Cultures of Various Human Tissue," *Journal of Immunology,* 1952;69;645-671.
Retrieved from Children of God for Life, "Forsaking God for the Sake of Science," https://cogforlife.org/2012/06/13/polioperversion/.

[3]Are Fetal Cells Used to Make Vaccines? Reviewed March 6, 2020. Accessed April 23, 2020__https://www.chop.edu/centers-programs/vaccine-education-center/video/are-fetal-cells-used-make-vaccines.

[4]"Thank You, President Trump for Upholding the Dignity and Sanctity of Life!" Children of God for Life, 6 June 2019, cogforlife.org/2019/06/06/thank-you-president-trump-for-upholding-the-dignity-and-sanctity-of-life/#more-18820.

[5]McKenna, Kyle Christopher. "Use of Aborted Fetal Tissue in Vaccines and Medical Research Obscures the Value of All Human Life." *The Linacre Quarterly,* SAGE Publications, Feb. 2018, www.ncbi.nlm.nih.gov/pmc/articles/PMC6027112/.

[6]Garry, D J, et al. "Are There Really Alternatives to the Use of Fetal Tissue from Elective Abortions in Transplantation Research?" *The New England Journal of Medicine,* U.S. National Library of Medicine, 26 Nov. 1992, www.ncbi.nlm.nih.gov/pubmed/1435888.

[7]O'Shea, *Vaccination is Not,* p. 167.

[8]Vaccine Ingredients – Fetal Tissues. Children's Hospital of Philadelphia https://www.chop.edu/centers-programs/vaccine-education-center/vaccine-ingredients/fetal-tissues. Reviewed February 27, 2020.

Chapter 2 Not Just Two, but Too Many to Count

[1]Dostoevsky, Fyodor. *The Brothers Karamazov: A Novel in Four Parts With Epilogue.* Farrar, Straus and Giroux, 2002. p. 242.

[2]Wadman, Meredith. "The Truth about Fetal Tissue Research." *Scientific American,* 9 Dec. 2015, www.scientificamerican.com/article/the-truth-about-fetal-tissue-research/.

[4]3Cell Collections A Publication of the Coriell Cell Repositories 2003/2004." Children of God for Life, Coriell Institute for Medical Research, 2004, cogforlife.org/wp-content/uploads/2012/05/imr90CoriellFullReport.pdf. p. 5.

[4]Ma, Bo, et al. "Characteristics and Viral Propagation Properties of a New Human Diploid Cell Line, Walvax-2, and Its Suitability as a Candidate Cell Substrate for Vaccine Production." *Human Vaccines & Immunotherapeutics,* Taylor & Francis, 2015, www.ncbi.nlm.nih.gov/pubmed/25803132.

[5]Knaup, G, et al. "Human Diploid Lung Fibroblast Cell Lines WI 26 and WI 38 Exhibit Isozyme Shift of Alkaline Phosphatase after Viral Transformation." *Clinica Chimica Acta; International Journal of Clinical Chemistry,* U.S. National Library of Medicine, 1 Sept. 1978, www.ncbi.nlm.nih.gov/pubmed/699332.

[6]Lorenzo, Elianet, et al. "Plasticity of the HEK-293 Cells, Related to the Culture Media, as Platform to Produce a Subunit Vaccine against Classical Swine Fever Virus." *AMB Express,* Springer Berlin Heidelberg, 5 Sept. 2019, www.ncbi.nlm.nih.gov/pmc/articles/PMC6728104/.

[7]VACCINES AND RELATED BIOLOGICAL PRODUCTS ADVISORY COMMITTEE MEETING." *Wayback.archive-It.org,* UNITED STATES OF AMERICA FOOD AND DRUG ADMINISTRATION CENTER FOR BIOLOGICS EVALUATION AND RESEARCH, 16 May 2001, wayback.archive-it.org/7993/20170404095417/https:/www.fda.gov/ohrms/dockets/ac/01/transcripts/3750t1_01.pdf.

[8]"NIH Guide: RESOURCES AVAILABLE FOR CONDUCTING RESEARCH ON AGING." *National Institutes of Health,* U.S. Department of Health and Human Services, grants.nih.gov/grants/guide/notice-files/not93-196.html.

[9]L. Hayflick, "The Limited In Vitro Lifetime of Human Diploid Cell Strains," *Experimental Cell Research,* Vol 37, pp. 614-636 1964. Retrieved from: https://cogforlife.org/Hayflick1964ExpCell.pdf.

[10]"Aborted Fetal Material Used in Vaccines and Medicines." Children of God for Life. https://cogforlife.org/wp-content/uploads/AbortedFetalCellLines.pdf.

[11]Humphries, Suzanne, and Bystrianyk, Roman. *Dissolving Illusions: Disease, Vaccines and the Forgotten History.* CreateSpace, 2015. p. 451.

[12]Baicus, A. "History of polio vaccination," *World Journal of Virology.* August 12, 2012. Date Accessed June 24, 2020. https://www.ncbi.nlm.nih.gov/pmc/articles/PMC3782271/.

[13]Children of God for Life, "Aborted Fetal Material Used in..."

[14]Ibid.

[15]L. Hayflick and P.S. Moorhead, *The Serial Cultivation of Human Diploid Cell Strains*, Experimental Cell Research Vol 25, pp 585-621, 1961. pp. 586-587.
Retrieved from: https://cogforlife.org/Hayflick1961ExpCell.pdf.

[16]Ibid., p. 591.

[17]Ibid., p. 585.

[18]Ibid., p. 618.

[19]Ibid., pp. 590-591.

[20]Ibid., p. 587.

[21]L. Hayflick, "The Limited In Vitro Lifetime…" p. 615.

[22]Plotkin et al, "Attenuation of RA 27/3 Rubella Virus in WI-38 Human Diploid Cell," *American Journal of Diseases of Children*, Vol. 118, pp. 178-179, August 1969. p. 178
Retrieved from: https://cogforlife.org/AmJDisChildRA273inWI-38.pdf.

[23]Hoskins, J.M. and Plotkin, S.A. "Behaviour of rubella virus in human diploid cell strains: I. Growth of Virus," *Archiv f Virusforschung*, Bd. 21, H. 3-4: 283-295, January 1967. pp. 284, 285, 293.
Retrieved from: https://cogforlife.org/PlotkinNumberOfAbortions.pdf.

[24]Ibid., p. 297.

[25]Jacobs et al, "Characteristics of a Human Diploid Cell Designated MRC-5," *Nature*, Vol 227, pp. 168 -170, 1970. p. 168.
Retrieved from: https://cogforlife.org/NatureJacobsMRC-5.pdf.

[26]Hoskins, "Behaviour of rubella virus in…" p. 297.

[27]Plotkin et al, "Studies of Immunization with living rubella virus; Trials in Children With a Strain Cultured From an Aborted Fetus," *American Journal Diseases in Children*, Vol 110, pp. 381-389, 1965. pp. 381, 382.
Retrieved from: https://cogforlife.org/AmJDisChildPlotkinRubellaVirus.pdf.

[28]G. Sven, S. Plotkin and K. McCarthy, "Gamma Globulin Prophylaxis; Inactivated Rubella Virus Production and Biological Control of Live Attenuated Rubella Virus Vaccines," *American Journal Diseases in Children*, Vol.. 118 Aug 1969. pp. 377-378.
Retrieved from: https://cogforlife.org/AmJDisChildMcCarthyGard.pdf.

[29]MATHESON VS. SCHMITT: DEPOSITION OF STANLEY A. PLOTKIN, M.D. CASE # 2015-831539-DM, JANUARY 11, 2018, COUNTY OF OAKLAND CIRCUIT COURT, FAMILY DIVISION, MICHIGAN.

[30]"Dr. Stanley 'Human in Form' Plotkin on using 76 fetuses (hearts, lungs, etc) to make vaccines," https://www.youtube.com/watch?v=bvBszdGBOxM&t=6s.

[31]"Dr. Stanley 'Human-in-Form' Plotkin, re: experiments on orphans/mentally retarded/prisoners," https://www.youtube.com/watch?v=yevV_slu7Dw&t=34s.

Chapter 3 God's Way to Avoid Disease

[1]*101 Scientific Facts & Foreknowledge.* Eternal Productions.
Content also available on http://www.eternal-productions.org/101science.html.

[2]Guyer, B., et al. "Annual Summary of Vital Statistics: Trends in the Health of Americans During the 20th Century." *Pediatrics*, vol. 106, no. 6, 2000, pp. 1307–1317. doi:10.1542/peds.106.6.1307. pp. 1314, 1315.

[3]James, Walene. *Immunization: The Reality behind the Myth*. Bergin & Garvey, 1995. p. 32.

[4]Phillips, Alan G. "Dispelling Vaccination Myths: An Introduction to the Contradictions Between Medical Science and Immunization Policy," Revised 2002, Myth #9 2007. p. 1.
PDF available: https://www.vacinfo.org/uploads/7/9/8/5/79856028/dispelling_vaccination_myths.pdf.

[5]Mendelsohn, Robert S. *How to Raise a Healthy Child... in Spite of Your Doctor*. Ballantine, 1987. p. 231.

[6]Scheibner, Viera. *Vaccination: 100 Years of Orthodox Research Shows That Vaccines Represent a Medical Assault on the Immune System*. V. Scheibner, 1993. p. 137.

[7]Humphries, *Dissolving Illusions*, p. 167.

[8]Habakus, Louise Kuo., et al. *Vaccine Epidemic: How Corporate Greed, Biased Science, and Coercive Government Threaten Our Human Rights, Our Health, and Our Children*. Skyhorse, 2012. p. 24.

[9]Ibid., pp. 27-28.

[10]Humphries, *Dissolving Illusions*, p. 342.

[11]Roser, Max. "Reduction of Cases and Deaths of Vaccine-Preventable Diseases in the US." Our World in Data, 19 Mar. 2015,
https://ourworldindata.org/reduction-of-cases-and-deaths-of-vaccine-preventable-diseases-in-the-us.

[12]Huff, Darrell, and Irving Geis. *How to Lie with Statistics*. W.W. Norton & Co., 2006. p. 100.

[13]Ibid., p. 86.

[14]Humphries, *Dissolving Illusions*, In Chapter 17 "Belief and Fear," some of Paul Offit's statements from his book *Deadly Choices—How the Anti-Vaccine Movement Threatens Us All* are compared to official statistics and the discrepancies are discussed, pp. 458-461.

[15]Ibid., p. 473.

[16]Scheibner, *Vaccination: 100 Years*, p. 29.

[17]Report of the Task Force on Pertussis and Pertussis Immunization—1988. James D. Cherry, Philip A. Brunell, Gerald S. Golden and David T. Karzon. Pediatrics June 1988, 81 (6) 933-984. p. 973.

[18]Wilder, Alexander. *The Fallacy of Vaccination*, The Metaphysical Publishing Co. 1899. p. 15
Available online: https://collections.nlm.nih.gov/ext/dw/101229606/PDF/101229606.pdf.

[19]Nguyen, Wilder. *Vaccination a Medical Fallacy*, EnCognitive.com, 2018. p. 142.

[20]Ibid., pp. 142, 143.

[21]Peebles, Vaccination, A Curse, p. 61.

[22]Boylston, Arthur. "The Origins of Inoculation." *Journal of the Royal Society of Medicine,* Royal Society of Medicine Press, July 2012, www.ncbi.nlm.nih.gov/pmc/articles/PMC3407399/.

[23]Riedel, S. "Edward Jenner and the history of smallpox and vaccination." Proc (Bayl Univ Med Cent). 2005 Jan. https://www.ncbi.nlm.nih.gov/pmc/articles/PMC1200696/.

[24]Peebles, Vaccination, A Curse, p. 14.

[25]Gunn, Robert A. *Vaccination: Its Fallacies and Evils (Classic Reprint).* Third ed., Facsimile Publisher, 2019. p. 5.

[26]Wolfe, Robert M, and Lisa K Sharp. "Anti-Vaccinationists Past and Present." *BMJ (Clinical Research Ed.),* BMJ, 24 Aug. 2002, www.ncbi.nlm.nih.gov/pmc/articles/PMC1123944/.

[27]Nguyen, Wilder, *Vaccination a Medical,* p. 142.

[28]McBean, Eleanor. *The Poisoned Needle: Suppressed Facts about Vaccination.* Health Research, 1993. p. 59.

[29]Gunn, *Vaccination: Its Fallacies,* p. 1.

[30]Wilder, *The Fallacy of,* p. 10.

[31]Humphries, *Dissolving Illusions,* pp. 95, 115.

[32]Ibid., p. 125.

[33]Ibid., p. 138.

[34]Scheibner, *Vaccination: 100 Years,* p. 208.

[35]Humphries, *Dissolving Illusions,* pp. 115-116.

[36]Gunn, *Vaccination: Its Fallacies,* pp. 1, 21.

[37]McBean, *The Poisoned Needle,* p. 25.

[38]Ibid., pp. 13-14.

[39]Ibid., p. 14.

[40]Peebles, *Vaccination, A Curse,* pp. 82-83.

[41]McBean, *The Poisoned Needle,* p. 30.

[42]Peebles, *Vaccination, A Curse,* pp. 84-85.

[43]McBean, *The Poisoned Needle,* pp. 35-38. (This last quote is also from John Baron's *The Life of Edward Jenner M.D. Vol. 2,* p. 374.)

[44]Ibid., p. 29.

[45]Baron, John. *The Life of Edward Jenner M.D. With Illustrations of His Doctrines, and Selections from His Correspondence. Volume 2.* Cambridge University Press. 2014. p. 304.

[46]McBean, *The Poisoned Needle,* p. 21.

[47]Huff, *How to Lie,* p. 11.

[48]Nguyen, Wilder. *Vaccination a Medical*, p. 154.

[49]Peebles, *Vaccination, A Curse*, p. 48.

[50]Ibid., p. 150.

[51]Theves, G. [Smallpox: An Historical Review]. *Bull Soc Sci Med Grand Duche Luxemb.* 1997. https://pubmed.ncbi.nlm.nih.gov/9303824/.

[52]Benedek, Thomas G. "Vaccination-induced Syphilis and the Hübner Malpractice Litigation." *Perspect Biol Med.* Winter 2012. https://pubmed.ncbi.nlm.nih.gov/22643719/.

[53]Peebles, *Vaccination, A Curse*, p. 28.

[54]Ibid., p. 118.

[55]Ibid., p. 54.

[56]McBean, *The Poisoned Needle*, p. 21.

[57]Peebles, *Vaccination, A Curse*, p. 40.

[58]Ibid., p. 84.

[59]Ibid., p. 87.

[60]Ibid., pp. 80-81.

[61]Ibid., pp. 125-126.

[62]Nguyen, Wilder. *Vaccination a Medical*, p. 171.

[63]Peebles, *Vaccination, A Curse*, p. 101.

[64]McBean, *The Poisoned Needle*, p. 59.

[65]Peebles, *Vaccination, A Curse*, p. 100.

[66]Ibid., pp. 173-174.

[67]Ibid., pp. 172-173.

[68]Ibid., p. 130.

[69]Ibid.

[70]Nguyen, Wilder. *Vaccination a Medical*, p. 201.

[71]Ibid., p. 189.

[72]Ibid., p. 173.

[73]Gunn, *Vaccination: Its Fallacies*, p. 25.

[74]McBean, *The Poisoned Needle*, pp. 33-34.

[75]Peebles, *Vaccination, A Curse*, p. 153.

[76]Wilder, *The Fallacy of*, p. 15.

[77]Ibid.

[78]Ibid., pp. 17-18.

[79]Nguyen, Wilder. *Vaccination a Medical*, p. 156.

[80]Wilder, *The Fallacy of*, pp. 13, 14.

[81]Ibid., p. 6.

[82]Peebles, Vaccination, A Curse, p. 317.

[83]Hak E; Schönbeck Y; De Melker H; Van Essen GA; Sanders EA; "Negative Attitude of Highly Educated Parents and Health Care Workers towards Future Vaccinations in the Dutch Childhood Vaccination Program." *Vaccine,* U.S. National Library of Medicine, 2005, pubmed.ncbi.nlm.nih.gov/15837208/.

[84]Wilder, *The Fallacy of,* p. 15.

[85]Peebles, Vaccination, A Curse, p. 325.

[86]Ibid., pp. 298-299.

[87]Ibid., p. 286.

[88]Wallace, Alfred Russell, *The Wonderful Century, its Successes and its Failures.* Scholar Select, 2015. p. 232.

[89]Peebles, *Vaccination, A Curse,* pp. 251-252.

[90]Ibid., p. 190.

[91]Wilder, *The Fallacy of,* p. 16.

[92]Peebles, *Vaccination, A Curse,* pp. 298, 299.

[93]Ibid., p. 313.

[94]Ibid., pp. 290-291.

[95]Ibid., pp. 61-62.

[96]Humphries, *Dissolving Illusions,* p. 124.

[97]Mullins, Eustace. *Murder by Injection: the Story of the Medical Conspiracy against America.* Omnia Veritas Ltd, 2016. p. 136.

[98]Coulter, Harris L., and Barbara Loe. Fisher. *A Shot in the Dark: Why the P in the DPT Vaccination May Be Hazardous to Your Child's Health.* Penguin, 1991. p. 197.

[99]Children of God for Life, "Aborted Fetal Material Used in…"

[100]Stern, Alexandra Minna. "Sterilized in the Name of Public Health: Race, Immigration, and Reproductive Control in Modern California." *American Journal of Public Health,* 2005, July 2005, www.ncbi.nlm.nih.gov/pmc/articles/PMC1449330/.

[101]Humphries, *Dissolving Illusions,* pp. 222-223.

[102]Maready, Forrest. *The Moth in the Iron Lung: a Biography of Polio.* Feels Like Fire, 2018. p. 42.

[103]Ibid.

[104]Ibid., p. 263.

[105]Ibid., p. 42.

[106]Ibid., pp. 80-81.

[107]Ibid., p. 37.

[108]Ibid., pp. 39-40.

[109]Ibid., pp. 37-38.

[110]Cathcart, Charles Sanderson, *Analyses of Paris green and lead arsenate*. New Jersey Agricultural Experiment Stations, 1909. p. 8.

[111]E. C. Seguin, M. D., "Myelitis Following Acute Arsenical Poisoning (By Paris or Schwein-Furth Green)," *Journal of Nervous and Mental Disease*, vol. IX, no. 4. October 1881. p. 9.

[112]Maready, *The Moth in*, p. 68.

[113]YouTube Video: "DDT so safe you can eat it 1947," https://www.youtube.com/watch?v=gtcXXbuR244.

[114]Maready, *The Moth in*, p. 54.

[115]Ibid., p. 10.

[116]McBean, The Poisoned Needle, p. 104.

[117]Ibid., p. 106.

[118]Ibid., p. 105.

[119]HIll, A B, and Knowelden, J. "Inoculation and Poliomyelitis; a Statistical Investigation in England and Wales in 1949." *British Medical Journal*, U.S. National Library of Medicine, 1 July 1950, www.ncbi.nlm.nih.gov/pmc/articles/PMC2038021/.

[120]Wyatt, H.V., et al. "Unnecessary Injections and Paralytic Poliomyelitis in India." *OUP Academic*, Oxford University Press, 1 Sept. 1992, academic.oup.com/trstmh/article-abstract/86/5/546/1923529?redirectedFrom=fulltext.

[121]"Disease Factsheet about Poliomyelitis." *European Centre for Disease Prevention and Control*, 11 July 2018, www.ecdc.europa.eu/en/poliomyelitis/facts.

[122]McBean, The Poisoned Needle, pp. 107, 109.

[123]Ibid., p. 107.

[124]Phillips, Vaccination Myths, p. 11.

[125]Hofstetter HH; Shive CL; Forsthuber TG. "Pertussis Toxin Modulates the Immune Response to Neuroantigens Injected in Incomplete Freund's Adjuvant: Induction of Th1 Cells and Experimental Autoimmune Encephalomyelitis in the Presence of High Frequencies of Th2 Cells." *Journal of Immunology (Baltimore, Md. : 1950)*, U.S. National Library of Medicine, 2002, pubmed.ncbi.nlm.nih.gov/12077236/.

[126]Emerson, Mitchell R, et al. "Enhancing the Ability of Experimental Autoimmune Encephalomyelitis to Serve as a More Rigorous Model of Multiple Sclerosis through Refinement of the Experimental Design." *Comparative Medicine*, American Association for Laboratory Animal Science, Apr. 2009, www.ncbi.nlm.nih.gov/pmc/articles/PMC2703151/.

[127]Gregg, Kelsey A, and Tod J Merkel. "Pertussis Toxin: A Key Component in Pertussis Vaccines?" *Toxins*, MDPI, 21 Sept. 2019, www.ncbi.nlm.nih.gov/pmc/articles/PMC6832755/.

[128]Coulter, Harris L. *Vaccination, Social Violence, and Criminality: the Medical Assault on the American Brain.* Atlantic, 1990. p. 102.

[129]Scheibner, *Vaccination 100 Years,* p. 151.

[130]Encephalopathies Following Prophylactic Pertussis Vaccine. Randolph K. Byers, Frederic C. Moll. *Pediatrics,* Apr 1948, 1 (4) 437-457. pp. 442, 443.

[131]Shannon, Sara. *Radiation Protective Foods: a Menu for the Nuclear Age.* AuthorHouse, 2012. p. 93.

[132]James, *Immunization: The Reality,* p. 120.

[133]Akizuki, Tatsuichirō, and Gordon Honeycombe. *Nagasaki 1945: the First Full-Length Eyewitness Account of the Atomic Bomb Attack on Nagasaki.* Quartet Books, 1982. pp. 111, 154.

[134]McCormick, W J. "POLIOMYELITIS: VITAMIN B DEFICIENCY A POSSIBLE FACTOR IN SUSCEPTIBILITY." *Canadian Medical Association Journal,* U.S. National Library of Medicine, Mar. 1938, www.ncbi.nlm.nih.gov/pmc/articles/PMC536391/. pp. 261-262.

[135]Ibid., p. 263.

[136]McBean, The Poisoned Needle, p. 104.

[137]Ibid., pp. 102, 104.

[138]Ibid., p. 109.

[139]Ibid.

[140]Ibid., p. 155.

[141]James, *Immunization: The Reality,* p. 54.

[142]Ibid., pp. 61-62.

[143]Maready, *Moth in the,* pp. 63-64.

[144]Ibid., p. 82.

[145]Ibid., p. 77.

[146]Humphries, *Dissolving Illusions,* p. 227.

[147]Maready, *Moth in the,* p. 95.

[148]Walene, *Immunization: The Reality,* pp. 35, 37.

[149]Ibid., pp. 36-37.

[150]Ibid., pp. 35, 36.

[151]Ibid., p. 37.

[152]United States, Congress, Cong. Senate, Labor and Human Resources, and Jonas Salk. *Hearing before the Committee on Labor and Human Resources, United States Senate,* Gov. Print. Off., 1984, pp. 169–173. 98th Congress, 2nd session, document Senate-Hrg-98-1060. p. 167.
Retrieved from: https://files.eric.ed.gov/fulltext/ED255480.pdf (quote on page 172 of this pdf).

[153]Mendelsohn, Robert. *Confessions of a Medical Heretic.* Contemporary Books Inc, 1990. p. 145.

[154]Maready, Moth in the, pp. 248, 266.

[155]Scheibner, *Vaccination: 100 Years,* p. 180.

[156]Mullins, *Murder by Injection,* p. 137.

[157]James, *Immunization: The Reality,* pp. 36, 37.

[158]Humphries, *Dissolving Illusions,* pp. 230-231.

[159]Scheibner, *Vaccination: 100 Years,* p. 261.

[160]Maready, *Moth in the,* p. 11.

[161]Ibid., p. 101.

[162]Maready, Forrest. *Unvaccinated: Why Growing Numbers of Parents Are Choosing Natural Immunity for Their Children.* Feels Like Fire, 2018. p. 32.

[163]Maready, *Moth in the,* p. 255.

[164]Ibid., p. 249.

[165]Ibid., p. 260.

[166]Ibid., pp. 260-261.

[167]"Maybe FDR Didn't Have Polio, Scientists Say." *CNN,* Cable News Network, 31 Oct. 2003, www.cnn.com/2003/HEALTH/10/31/roosevelt.polio.reut/.

[168]Levy, Thomas E. *Curing the Incurable: Vitamin C, Infectious Diseases, and Toxins.* MedFox Publishing, LLC, 2011. pp. 53-54, 55.

[169]Ibid., p. 75.

[170]Scheibner, *Vaccination: 100 Years,* p. xxi.

[171]Humphries, *Dissolving Illusions,* p. 337.

[172]Scheibner, *Vaccination: 100 Years,* pp. xix-xx.

[173]Ewing, Graham E. "What Is Regressive Autism and Why Does It Occur? Is It the Consequence of Multi-Systemic Dysfunction Affecting the Elimination of Heavy Metals and the Ability to Regulate Neural Temperature?" *North American Journal of Medical Sciences,* Medknow Publications & Media Pvt Ltd, July 2009, www.ncbi.nlm.nih.gov/pmc/articles/PMC3364648/.

[174]Markowitz, L E, et al. "Patterns of Transmission in Measles Outbreaks in the United States, 1985-1986." *The New England Journal of Medicine,* U.S. National Library of Medicine, 12 Jan. 1989, www.ncbi.nlm.nih.gov/pubmed/2911293.

[175]Poland GA; Jacobson RM, "Failure to Reach the Goal of Measles Elimination. Apparent Paradox of Measles Infections in Immunized Persons." *Archives of Internal Medicine,* U.S. National Library of Medicine, 1994, pubmed.ncbi.nlm.nih.gov/8053748/.

[176]Peebles, Vaccination, A Curse, p. 94.

[177]Walene, *Immunization: The Reality,* p. 14.

[178]Mendelsohn, *How to Raise*, p. 238.

[179]Hoption Cann SA; Van Netten JP; Van Netten C; "Acute Infections as a Means of Cancer Prevention: Opposing Effects to Chronic Infections?" *Cancer Detection and Prevention,* U.S. National Library of Medicine, 2006, pubmed.ncbi.nlm.nih.gov/16490323/.

[180]Albonico HU; Bräker HU; Hüsler J. "Febrile Infectious Childhood Diseases in the History of Cancer Patients and Matched Controls." *Medical Hypotheses,* U.S. National Library of Medicine, 1998, pubmed.ncbi.nlm.nih.gov/9824838/.

[181]Mendelsohn, *How to Raise*, p. 233.

[182]Levy, *Curing the Incurable*, pp. 68-69.

[183]Ewing, "What Is Regressive Autism and…?"

[184]Pollock, T M, et al. "Severity of Whooping Cough in England before and after the Decline in Pertussis Immunisation." *Archives of Disease in Childhood,* U.S. National Library of Medicine, Feb. 1984, www.ncbi.nlm.nih.gov/pmc/articles/PMC1628449/.

[185]Scheibner, *Vaccination 100 Years*, p. 33.

[186]Ibid., p. 49.

[187]Hutchins SS; Cochi SL; Brink EW; Patriarca PA; Wassilak SG; Rovira EZ; Hinman AR; "Current Epidemiology of Pertussis in the United States." *The Tokai Journal of Experimental and Clinical Medicine,* U.S. National Library of Medicine, 1988, pubmed.ncbi.nlm.nih.gov/2856218/.

[188]Gill, Christopher, et al. "The Relationship between Mucosal Immunity, Nasopharyngeal Carriage, Asymptomatic Transmission and the Resurgence of *Bordetella Pertussis.*" *F1000Research,* 25 Aug. 2017, www.ncbi.nlm.nih.gov/pmc/articles/PMC5580413/.

[189]*Salma Abdalla,* et al. "Resurgence of Whooping Cough May Owe to Vaccine's Inability to Prevent Infections: SPH: Boston University." *School of Public Health,* www.bu.edu/sph/2017/09/21/resurgence-of-whooping-cough-may-owe-to-vaccines-inability-to-prevent-infections/.

[190]Ibid.

[191]Humphries, *Dissolving Illusions*, pp. 330-331.

[192]Ibid., p. 479.

[193]O'Shea, *Vaccination is Not*, p. v.

[194]Phillips, Vaccination Myths, p. 6..

[195]O'Shea, *Vaccination is Not*, pp. 59, 60

[196]Ibid., pp. 63-64.

[197]Peebles, Vaccination, A Curse, p. 93.

[198]Nguyen T., McBean E., Martson S., Honorof I., *Vaccines: The Biggest Medical Fraud in History,* EnCognitive.com, 2018, p. 36.

[199]Habakus, *Vaccine Epidemic*, pp. 127-128.

[200]Levy, *Curing the Incurable.*

[201]Tim O'Shea, *Vaccination is Not,* p. 14.

[202]Ibid., pp. 15, 16.

[203]Ralph R. Tompsett (1996) The Discovery of Penicillin, Baylor University Medical Center Proceedings, 9:3, 37-38, DOI: 10.1080/08998280.1996.11929983. p. 38.

Chapter 4 Many Possible Doubts

[1]Walene, *Immunization: The Reality,* p. 9.

[2]Mendelsohn, *How to Raise,* p. 230.

[3]McBean, The Poisoned Needle, pp. 104-105.

[4]Mendelsohn, *How to Raise,* p. 232.

[5]United States Federal Register, *Federal Register Volume 49, Issue 107,* Office of the Federal Register, National Archives and Records Administration, 1984, pp. 22751–23018. p. 23007.
Retrieved from: https://tile.loc.gov/storage-services/service/ll/fedreg/fr049/fr049107/fr049107.pdf.

[6]Habakus, *Vaccine Epidemic,* pp. 2, 5.

[7]Ibid., p. 15.

[8]SUPREME COURT OF THE UNITED STATES. *BRUESEWITZ ET AL. v. WYETH LLC, FKA WYETH, INC., ET AL.* www.supremecourt.gov/opinions/10pdf/09-152.pdf. (Search "Comment k provides as follows.")

[9]Habakus, *Vaccine Epidemic,* p. 19.

[10]Inbar, Rotem, et al. "Behavioral Abnormalities in Female Mice Following Administration of Aluminum Adjuvants and the Human Papillomavirus (HPV) Vaccine Gardasil." *Immunologic Research,* U.S. National Library of Medicine, Feb. 2017, www.ncbi.nlm.nih.gov/pubmed/27421722.

[11]Exley, Christopher. "Aluminium-Based Adjuvants Should Not Be Used as Placebos in Clinical Trials." *Vaccine,* vol. 29, 2011, Letter to the Editor doi:10.1016/j.vaccine.2011.08.062.

[12]Xu DZ, Zhao K, Guo LM, Chen XY, Wang HF, Zhang JM, et al. "A randomised controlled phase IIb trial of antigen–antibody immunogenic complex therapeutic vaccine in chronic hepatitis B patients." *PLoS ONE.* https://www.ncbi.nlm.nih.gov/pmc/articles/PMC2430617/.

[13]Bhatla N, Suri V, Basu P, Shastri S, Datta S K,Bi D,et al. "Immunogenicity and safety of human papillomavirus-16/18 AS04-adjuvanted cervical cancer vaccine in healthy Indian women." *J Obstet Gynaecol Res* 2010;36:123–32. https://pubmed.ncbi.nlm.nih.gov/20178538/.

[14]Landry N, Ward BJ, Trepanier S, Montomoli E, Dargis M, Lapini G, et al. "Preclinical and clinical development of plant-made virus-like particle vaccine against avian

H5N1 influenza." *PLoS ONE* 2010;5:e15559. https://pubmed.ncbi.nlm.nih.gov/21203523/.

[15]Wang X Y, Zhang X X, Yao X, Jiang J H, Xie Y H, Yuan Z H, et al. "Serum HbeAgsero-conversion correlated with decrease of HbsAg and HBV DNA in chronic hepatitis B patients treated with a therapeutic vaccine." *Vaccine* 2010;28:8169–74. https://pubmed.ncbi.nlm.nih.gov/20937312/.

[16]Garnock-Jones KP, Giuliano AR. "Quadrivalent human papillomavirus (HPV) types 6, 11, 16, 18 vaccine for the prevention of genital warts in males." *Drugs* 2011;71:591–602. https://pubmed.ncbi.nlm.nih.gov/21443282/.

[17]Kim SC, Song YS, Kim YT, Kim YT, Ryu KS, Gunapalaiah B, et al. "Human papillomavirus 16/18 AS04-adjuvanted cervical cancer vaccine: immunogenicity and safety in 15–25 years old healthy Korean women." *J Gynecol Oncol* 2011;22:67–75. https://www.ncbi.nlm.nih.gov/pmc/articles/PMC3152758/.

[18]Shaw, C A, and L Tomljenovic. "Aluminum in the Central Nervous System (CNS): Toxicity in Humans and Animals, Vaccine Adjuvants, and Autoimmunity." *Immunologic Research,* U.S. National Library of Medicine, July 2013, www.ncbi.nlm.nih.gov/pubmed/23609067.

[19]Maready, Forrest. *Crooked: Man-Made Disease Explained ; the Incredible Story of Metal, Microbes, and Medicine - Hidden within Our Faces.* Feels Like Fire, 2018. p. 103.

[20]Crépeaux G; Eidi H; David MO; Baba-Amer Y; Tzavara E; Giros B; Authier FJ; Exley C; Shaw CA; Cadusseau J; Gherardi RK; "Non-Linear Dose-Response of Aluminium Hydroxide Adjuvant Particles: Selective Low Dose Neurotoxicity." *Toxicology,* U.S. National Library of Medicine, 2017, pubmed.ncbi.nlm.nih.gov/27908630/.

[21]Maready, *Crooked: Man-Made,* p. 104.

[22]Ibid., pp. 107, 116.

[23]Wacholder S, Chen B E, Wilcox A, Macones G, Gonzalez P, Befano B, et al. "Risk of miscarriage with bivalent vaccine against human papillomavirus (HPV) types 16 and 18: pooled analysis of two ramdomised controlled trials." *British Med J* 2010;340:c712. https://pubmed.ncbi.nlm.nih.gov/20197322/.

[24]Medina DMR, Valencia A, de Velasquez A, Huang LM, Prymula R, García- Sicilia J, et al. "Safety and immunogenicity of the HPV-16/18 AS04-adjuvanted vaccine: a randomised, controlled trial in adolescent girls." *J Adolesc Health* 2010;46:414–21. https://pubmed.ncbi.nlm.nih.gov/20413076/.

[25]Golomb, Beatrice A, et al. "What's in Placebos: Who Knows? Analysis of Randomized, Controlled Trials." *Annals of Internal Medicine,* U.S. National Library of Medicine, 19 Oct. 2010, www.ncbi.nlm.nih.gov/pubmed/20956710.

[26]"Double-Blind, Placebo-Controlled Study." Winchester Hospital, www.winchesterhospital.org/health-library/article?id=21849.

[27]Ewing, "What Is Regressive Autism and…?"

[28]Maready, *Unvaccinated: Why Growing,* p. 62.

[29]Davis, Garth, and Howard Jacobson. *Proteinaholic: How Our Obsession with Meat Is Killing Us and What We Can Do about It*. HarperOne, an Imprint of HarperCollins Publishers, 2016. p. 18.

[30]Ioannidis, John P. A. "Why Most Published Research Findings Are False." *PLOS Medicine,* Public Library of Science, 30 Aug. 2005, journals.plos.org/plosmedicine/article?id=10.1371%2Fjournal.pmed.0020124.

[31]George, Stephen L, and Marc Buyse. "Data Fraud in Clinical Trials." *Clinical Investigation,* U.S. National Library of Medicine, 2015, www.ncbi.nlm.nih.gov/pmc/articles/PMC4340084/.

[32]Coulter, *A Shot in,* p. 118.

[33]O'Shea, *Vaccination is Not,* p. 36.

[34]Barry, Kevin, et al. *Vaccine Whistleblower: Exposing Autism Research Fraud at the CDC.* Skyhorse Publishing, 2017. pp. xi-xii.

[35]Zheteyeva, Yenlik A, et al. "Adverse Event Reports after Tetanus Toxoid, Reduced Diphtheria Toxoid, and Acellular Pertussis Vaccines in Pregnant Women." *American Journal of Obstetrics and Gynecology,* U.S. National Library of Medicine, July 2012, www.ncbi.nlm.nih.gov/pubmed/22727350.

[36]Gill, "The Relationship between Mucosal Immunity..."

[37]Zheteyeva, "Adverse Event Reports after Tetanus Toxoid…"

[38]Goldman, G S. "Comparison of VAERS Fetal-Loss Reports during Three Consecutive Influenza Seasons: Was There a Synergistic Fetal Toxicity Associated with the Two-Vaccine 2009/2010 Season?" *Human & Experimental Toxicology,* SAGE Publications, May 2013, www.ncbi.nlm.nih.gov/pmc/articles/PMC3888271/.

[39]Habakus, *Vaccine Epidemic,* pp. 209-210.

[40]Ibid., p. 270.

[41]"Pediarix Vaccine Questions and Answers for Healthcare Providers | CDC." *Centers for Disease Control and Prevention,* Centers for Disease Control and Prevention, 2 Jan. 2013, www.cdc.gov/vaccines/vpd/hepb/hcp/faqs-hcp-pediarix.html.

[42]Ewing, "What Is Regressive Autism and…?"

[43]Thompson, William W., et al. "Early Thimerosal Exposure and Neuropsychological Outcomes at 7 to 10 Years." *New England Journal of Medicine,* 27 Sept. 2007, www.nejm.org/doi/full/10.1056/nejmoa071434.

[44]Barry, *Vaccine Whistleblower,* pp. xv-xvi.

[45]Coulter, *Vaccination, Social Violence,* pp. 143-144.

[46]Offit, Paul A., and Charlotte A. Moser. *Vaccine Safety and Your Child,* excerpted from *Vaccines & Your Child: Separating Fact from Fiction*. Columbia University Press, 2011. p. 19.

[47]Miller, Neil Z, and Gary S Goldman. "Infant Mortality Rates Regressed against Number of Vaccine Doses Routinely given: Is There a Biochemical or Synergistic

Toxicity?" *Human & Experimental Toxicology*, SAGE Publications, Sept. 2011, www.ncbi.nlm.nih.gov/pmc/articles/PMC3170075/.

[48]Stevenson, Heidi. "The Only Credible Cause of Sudden Infant Death Syndrome Is Vaccines." *Semantic Scholar*, 15 July 2012, pdfs.semanticscholar.org/f838/c63a827b28d6ef3a57bebda7c657ccd7df7b.pdf.

[49]Miller, "Infant Mortality Rates Regressed against…"

[50]Coulter, *A Shot in*, p. 53.

[51]Soldatenkova, Valentina A., and F. Edward Yazbak. "An Investigation of Infant Deaths Following Initial Hepatitis B Vaccination Based on the Vaccine Adverse Event Reporting System (VAERS), 1992-2002." *Semantic Scholar*, 2007, www.semanticscholar.org/paper/An-investigation-of-infant-deaths-following-initial-Soldatenkova-Yazbak/d477af0d2fd55b2c6bf6cc29bbb512f18139a225. p. 1418. (Hit the down arrow and select the vaclib.org option to see the full study.)

[52]Institute of Medicine (US) Committee to Review the Adverse Consequences of Pertussis and Rubella Vaccines. "Evidence Concerning Pertussis Vaccines and Deaths Classified as Sudden Infant Death Syndrome." *Adverse Effects of Pertussis and Rubella Vaccines: A Report of the Committee to Review the Adverse Consequences of Pertussis and Rubella Vaccines.*, U.S. National Library of Medicine, 1 Jan. 1991, www.ncbi.nlm.nih.gov/books/NBK234368/.

[53]Cherry, "Report of the Task Force…" p. 973.

[54]Coulter, *A Shot in*, p. 26.

[55]Scheibner, *Vaccination: 100 Years*, p. 63.

[56]Habakus, *Vaccine Epidemic*, p. 30.

[57]Soldatenkova, "An Investigation of Infant Deaths…"

[58]"Vaccination The Hidden Truth." *Vaccination MUST SEE! - The Hidden Truth - Australian Documentary (FULL)*, 1998, www.youtube.com/watch?v=P9CGfYPXDKg.

[59]Scheibner, *Vaccination: 100 years*, p. 10.

[60]Miller, Neil Z. *Miller's Review of Critical Vaccine Studies: 400 Important Scientific Papers Summarized for Parents and Researchers*. Page 16. New Atlantean Press, 2016. p. 16.

[61]Miller, "Infant Mortality Rates Regressed against..."

[62]Ibid.

[63]United States, Congress, Cong. House, COMMITTEE ON GOVERNMENT REFORM. "FACA: CONFLICTS OF INTEREST AND VACCINE DEVELOPMENT--PRESERVING THE INTEGRITY OF THE PROCESS." *FACA: CONFLICTS OF INTEREST AND VACCINE DEVELOPMENT--PRESERVING THE INTEGRITY OF THE PROCESS*, U.S. GOVERNMENT PRINTING OFFICE, 2000. 106th Congress, 2nd session, document 106-239.
https://www.govinfo.gov/content/pkg/CHRG-106hhrg73042/html/CHRG-106hhrg73042.htm

"It [Rotavirus Vaccine] was approved for use by the FDA in August 1998. It was recommended for universal use by the CDC in March 1999. Serious problems cropped shortly after it was introduced. Children started developing serious bowel obstructions. The vaccine was pulled from the U.S. market in October 1999. So the question is, was there evidence to indicate that the vaccine was not safe, and if so, why was it licensed in the first place? How good a job did the advisory committees do? We reviewed the minutes of the meetings. At the FDA's committee, there were discussions about adverse events. They were aware of potential problems. Five children out of 10,000 developed bowel obstructions. There were also concerns about children failing to thrive and developing high fevers, which as we know from other vaccine hearings, can lead to brain injury. Even with all of these concerns, the committee voted unanimously to approve it. At the CDC's committee, there was a lot of discussion about whether the benefits of the vaccine really justified the cost. Even though the cost benefit ratio was questioned, the committee voted unanimously to approve it. … Doctors place trust in the FDA and assume that if the FDA has licensed a drug, it's safe for use. Has that trust been violated? How confident in the safety and need of specific vaccines would doctors and parents be if they learned the following: One, that members, including the chair of the FDA and CDC advisory committees who make these decisions own stock in drug companies that make the vaccines. Two, that individuals on both advisory committees own patents for vaccines under consideration, or affected by the decisions of the committees. Three, that three out of the five of the members of the FDA's advisory committee who voted for the rotavirus vaccine had conflicts of interest that were waived. … Five, that the CDC grants conflict of interest waivers to every member of their advisory committee a year at a time, and allows full participation in the discussions leading up to a vote by every member, whether they have a financial stake in the decision or not. So they're discussing it, influencing other members possibly, whether they have a financial stake or not. … How confident can we be in the process when we learned that most of the work of the CDC advisory committee is done in ``working groups'' that meet behind closed doors, out of the public eye? Members who can't vote in the full committee because of conflicts of interest are allowed to work on the same issues in working groups, and there is no public scrutiny. I was appalled to learn that at least 6 of the 10 individuals who participated in the working group for the rotavirus vaccine had financial ties to pharmaceutical companies developing rotavirus vaccines. How confident can we be in the recommendations for the Food and Drug Administration when the chairman and other individuals on their advisory committee own stock in major manufacturers of vaccines?"

[64]"Birth-18 Years Immunization Schedule." *Centers for Disease Control and Prevention*, Centers for Disease Control and Prevention, 3 Feb. 2020, www.cdc.gov/vaccines/schedules/hcp/imz/child-adolescent.html.

[65]Habakus, *Vaccine Epidemic*, p. 309.

[66]*Children's Hospital of Philadelphia*, "Vaccine History: Developments by Year." The Children's Hospital of Philadelphia, 20 Nov. 2014, https://www.chop.edu/centers-programs/vaccine-education-center/vaccine-history/developments-by-year.

[67]Von Kries, Rüdiger, et al. "Sudden and Unexpected Deaths after the Administration of Hexavalent Vaccines (Diphtheria, Tetanus, Pertussis, Poliomyelitis, Hepatitis B, Haemophilius Influenzae Type b): Is There a Signal?" *European Journal of Pediatrics*, U.S. National Library of Medicine, Feb. 2005, www.ncbi.nlm.nih.gov/pubmed/15602672.

[68]Traversa, Giuseppe, et al. "Sudden Unexpected Deaths and Vaccinations during the First Two Years of Life in Italy: A Case Series Study." *PLOS ONE*, Public Library of Science, 26 Jan. 2011, journals.plos.org/plosone/article?id=10.1371%2Fjournal.pone.0016363.

[69]Kuhnert, Ronny, et al. "A Modified Self-Controlled Case Series Method to Examine Association between Multidose Vaccinations and Death." *Statistics in Medicine*, U.S. National Library of Medicine, 15 Mar. 2011, www.ncbi.nlm.nih.gov/pubmed/21337361.

[70]Barry, *Vaccine Whistleblower*, p. 109.

[71]Kalokerinos, Archie. *Every Second Child*. Keats Pub., 1981. pp. 67-68.

[72]Soldatenkova, "An Investigation of Infant Deaths..." p. 1417.

[73]Kalokerinos, *Every Second Child*, pp. 139-140.

[74]Scheibner, *Vaccination: 100 Years*, pp. 234, 235.

[75]Coulter, *A Shot in*, p. 46.

[76]Scheibner, *Vaccination: 100 Years*, p. 50.

[77]Ibid., p. 258.

[78]Coulter, *A Shot in*, p. 57.

[79]Scheibner, *Vaccination: 100 Years*, p. 258.

[80]Coulter, *A Shot in*, p. 33.

[81]Scheibner, *Vaccination: 100 Years*, p. 244.

[82]Ibid., pp. 107, 259, 261-262.

[83]Ibid., p. 59.

[84]Levy, Thomas E. *Hidden Epidemic: Silent Oral Infections Cause Most Heart Attacks and Breast Cancers*. MedFox Publishing, 2017. p. 59.

[85]Coulter, *A Shot in*, p. 51.

[86]James, Immunization: The Reality, p. 22.

[87]Coulter, *Vaccination, Social Violence*, pp. 23, 24.

[88]Habakus, *Vaccine Epidemic*, pp. 219, 225-226.

[89]Barry, *Vaccine Whistleblower*, pp. xvi-xvii, xviii, xix.

[90]Ibid., pp. 108-109.

[91]Ibid., p. 110.

[92]Ibid., p. xix.

[93]Ibid., p. xvii .

[94]Chhawchharia, Ritika, and Jacob M Puliyel. "Commentary--Controversies Surrounding Mercury in Vaccines: Autism Denial as Impediment to Universal Immunisation." *Indian Journal of Medical Ethics,* U.S. National Library of Medicine, 2014, www.ncbi.nlm.nih.gov/pubmed/25377033.

[95]Hviid A, et al. "Measles, Mumps, Rubella Vaccination and Autism: A Nationwide Cohort Study." *Annals of Internal Medicine,* 2019, www.acpjournals.org/doi/10.7326/M18-2101.

[96]Ibid.

[97]Hooker, Brian S., et al. "A Scientist's Rebuttal to the Danish Cohort Study." *Focus for Health,* 6 Sept. 2019, www.focusforhealth.org/scientists-rebuttal-to-danish-cohort-study/.

[98]*Federal Register Volume 49, Issue 107,* p. 23007.

[99]Habakus, *Vaccine Epidemic,* p. 8.

[100]Shaw, "Aluminum in the Central Nervous System..."

[101]Delong, Gayle. "A Positive Association Found between Autism Prevalence and Childhood Vaccination Uptake across the U.S. Population." *Journal of Toxicology and Environmental Health. Part A,* U.S. National Library of Medicine, 2011, www.ncbi.nlm.nih.gov/pubmed/21623535.

[102]Ewing, "What Is Regressive Autism and…?"

[103]Tomljenovic, Lucija, and Christopher A Shaw. "Do Aluminum Vaccine Adjuvants Contribute to the Rising Prevalence of Autism?" *Journal of Inorganic Biochemistry,* U.S. National Library of Medicine, Nov. 2011, www.ncbi.nlm.nih.gov/pubmed/22099159.

[104]Ewing, "What Is Regressive Autism and…?"

[105]Habakus, *Vaccine Epidemic,* pp. 46, 47.

[106]Mendelsohn, *Healthy Child,* p. 231.

[107]Vaccine Science & Options. Sound Choice Pharmaceutical Institute. www.soundchoice.org/ethical-vaccines/.

[108]Oehen S, Hengartner H, Zinkernagel RM. "Vaccination for disease." Science. 1991 Jan 11;251(4990):195-8. doi: 10.1126/science.1824801. PMID: 1824801. https://pubmed.ncbi.nlm.nih.gov/1824801/.

[109]Ewing, "What Is Regressive Autism and…?"

[110]Walene, *Immunization: The Reality,* pp. 9-10.

[111]Scheibner, *Vaccination: 100 Years,* p. 174.

[112]Manzotti, Francesca, et al. "Partial Third Nerve Palsy after Measles Mumps Rubella Vaccination." *Italian Journal of Pediatrics,* BioMed Central, 10 Sept. 2010, www.ncbi.nlm.nih.gov/pmc/articles/PMC2944153/.

[113]Scheibner, *Vaccination: 100 Years*, p. 257.

[114]Ibid., p. 241.

[115]Walene, *Immunization: The Reality*, pp. 12, 18.

[116]Coulter, *A Shot in*, p. 68.

[117]Walene, *Immunization: The Reality*, p. 19.

[118]Ibid.

[119]Coulter, *A Shot in*, p. 30.

[120]Ibid., p. 31.

[121]Walene, *Immunization: The Reality*, p. 13.

[122]Coulter, *A Shot in*, p. 50.

[123]Coulter, *Vaccination, Social Violence*, pp. 111-112.

[124]Habakus, *Vaccine Epidemic*, p. 169.

[125]Trebing, William P. *Good-Bye Germ Theory: Ending a Century of Medical Fraud and How to Protect Your Family*. Xlibris Corp., 2006. p. 190.

[126]Habakus, *Vaccine Epidemic*, p. 290.

[127]"Article Claiming Vaccines Cause Autoimmunity and Autism Due to Fetal DNA Contaminants Found Unsupported and Implausible." *Health Feedback*, 28 June 2019, healthfeedback.org/evaluation/article-claiming-vaccines-cause-autoimmunity-and-autism-due-to-fetal-dna-contaminants-found-unsupported-and-implausible-theresa-deisher/.

[128]Ewing, "What Is Regressive Autism and…?"

[129]Stewart, G T. "Vaccination against Whooping-Cough. Efficacy versus Risks." *Lancet (London, England)*, U.S. National Library of Medicine, 29 Jan. 1977, www.ncbi.nlm.nih.gov/pubmed/64761.

[130]Habakus, *Vaccine Epidemic*, p. 2.

[131]O'Shea, *Vaccination is Not*, p. 118.

[132]Habakus, *Vaccine Epidemic*, p. 4.

[133]Ibid., pp. 32, 48.

[134]Coulter, *A Shot in*, pp. 2-3.

[135]United States District Court Southern District of New York. *Informed Consent Action Network against The United States Department of Health and Human Services*. Case 1:18-cv-03215-JMF, 6 July 2018, www.icandecide.org/wp-content/uploads/2019/09/Stipulated-Order-copy-1.pdf.

[136]Ibid.

[137]Barry, *Vaccine Whistleblower*, pp. 106, 108.

[138]United States Federal Register, *Federal Register Volume 49, Issue 107*, pp. 23004, 23005.

[139]Greger, Michael. *How Not To Diet: Discover the Groundbreaking Science of Weight Loss*. Flatiron Books, 2019. p. 1.

[140]Mendelsohn, *Medical Heretic*, pp. 80, 129.

[141]Ibid., p. 16.

[142]Ibid., p. 19—"Doctors in general should be treated with about the same degree of trust as used car salesmen."

p. 163—"Experts' opinions should be considered worthless until proved otherwise by the strongest possible evidence."

[143]Humphries, *Dissolving Illusions*, p. 336.

[144]Ibid., p. 479.

[145]Habakus, *Vaccine Epidemic*, p. 6.

[146]Ibid., p. 273.

[147]Ibid., p. 129.

[148]Ibid., p. 158.

[149]O'Shea, *Vaccination is Not*, p. 193.

Chapter 5 Vaccinated vs. Unvaccinated Pilot Study

[1]Mawson, A et al. "Pilot Comparative Study on the Health of Vaccinated and Unvaccinated 6- to 12- Year Old U.S. Children." Journal of Translational Science, 2017, www.oatext.com/Pilot-comparative-study-on-the-health-of-vaccinated-and-unvaccinated-6-to-12-year-old-U-S-children.php.

[2]Ibid.

[3]Mawson, A. et al. "Preterm Birth, Vaccination and Neurodevelopmental Disorders: a Cross-Sectional Study of 6- to 12-Year-Old Vaccinated and Unvaccinated Children." *Oat Open Access Text* , Journal of Translational Science, 2017, www.oatext.com/Preterm-birth-vaccination-and-neurodevelopmental-disorders-a-cross-sectional-study-of-6-to-12-year-old-vaccinated-and-unvaccinated-children.php#gsc.tab=0.

(From what I understand, it has been inaccurately reported that studies 1 and 3 were both retracted.)

[4]Roland W. Sutter, MD. "Pertussis Hospitalizations and Mortality in the United States, 1985-1988." *JAMA*, American Medical Association, 15 Jan. 1992, jamanetwork.com/journals/jama/article-abstract/394566.

[5]Scheibner, *Vaccination: 100 Years*, p. 113.

[6]Mawson, "Pilot Comparative Study on the…"

[7]Maloney, Carolyn B. "H.R.2832 - 110th Congress (2007-2008): Comprehensive Comparative Study of Vaccinated and Unvaccinated Populations Act of 2007." *Congress.gov*, 22 June 2007, www.congress.gov/bill/110th-congress/house-bill/2832?s=1&r=44.

[8]Habakus, *Vaccine Epidemic*, p. 78.

[9]Phillips, Dispelling Vaccination Myths, pp. 8, 11.

[10]Skinner, Erik, and Alise Garcia. "States With Religious and Philosophical Exemptions From School Immunization Requirements," www.ncsl.org/research/health/school-immunization-exemption-state-laws.aspx.

[11]Offit, *Vaccine Safety and Your Child,* p. 9.

[12]Scheibner, *Vaccination: 100 Years,* p. 263.

[13]Coulter, *A Shot in,* p. 218.

[14]Habakus, *Vaccine Epidemic,* p. 93.

[15]Alfred Russel Wallace, *To Members of Parliament and Others Forty-Five Years of Registration Statistics, Proving Vaccination to be Both Useless and Dangerous,* 2013 Read Book Ltd. pp. 34-35.

[16]Habakus, *Vaccine Epidemic,* p. 261.

[17]McBean, The Poisoned Needle, pp. 85-86.

Chapter 6 Wait and Pray

[1]Schreibner, *100 Years,* p. 7.

[2]Ibid.

[3]Ibid.

[4]James, *Immunization: The Reality,* p. 150.

[5]Lao, Terence T. "Immune Persistence after Hepatitis B Vaccination in Infancy - Fact or Fancy?" *Human Vaccines & Immunotherapeutics,* Taylor & Francis, 3 May 2016, www.ncbi.nlm.nih.gov/pmc/articles/PMC4963055/.

[6]Lu CY; Chiang BL; Chi WK; Chang MH; Ni YH; Hsu HM; Twu SJ; Su IJ; Huang LM; Lee CY; "Waning Immunity to Plasma-Derived Hepatitis B Vaccine and the Need for Boosters 15 Years After Neonatal Vaccination." *Hepatology (Baltimore, Md.),* U.S. National Library of Medicine, 2004, pubmed.ncbi.nlm.nih.gov/15565627/.

[7]Trebing, *Good-Bye Germ,* pp. 181-182.

[8]Bernstein, Steve, and Michael Klompas. "Electronic Support for Public Health–Vaccine Adverse Event Reporting System (ESP:VAERS)." *Agency for Healthcare Research and Quality,* Harvard Pilgrim Health Care, Inc., 2010, digital.ahrq.gov/sites/default/files/docs/publication/r18hs017045-lazarus-final-report-2011.pdf.

[9]DOCTORS CALL FOR MORATORIUM ON HEPATITIS B VACCINE FOR SCHOOLCHILDREN CITING POTENTIAL DEADLY OUTCOMES. Association of American Physicians and Surgeons, Inc. July 8 1999. https://www.aapsonline.org/press/hepbpre.htm.

[10]Lyons-Weiler, James, and Robert Ricketson. "Reconsideration of the Immunotherapeutic Pediatric Safe Dose Levels of Aluminum." *Journal of Trace Elements in Medicine and Biology,* Urban & Fischer, 8 Mar. 2018, www.sciencedirect.com/science/article/pii/S0946672X17300950?via=ihub. (Quote is

not from the actual study but from the highlights of the study shown at the above URL page.)

[11]U. S. Food and Drug Administration. "CFR - Code of Federal Regulations Title 21." Volume 7, Chapter 1, Subchapter F, Part 610, Subpart B, Section 610.15. *Accessdata.fda.gov,* Last Updated April 1, 2019, www.accessdata.fda.gov/scripts/cdrh/cfdocs/cfcfr/CFRSearch.cfm?fr=610.15.

[12]Lyons-Weiler, "Reconsideration of the Immunotherapeutic Pediatric…"

[13]U.S. Food and Drug Administration. "CFR - Code of Federal Regulations Title 21." Volume 4. Chapter 1, Subchapter C, Part 201, Subpart G, Section 201.323. *Accessdata.fda.gov,* , Last Updated September 19, 2019, www.accessdata.fda.gov/scripts/cdrh/cfdocs/cfcfr/cfrsearch.cfm?fr=201.323.

[14]The Children's Hospital of Philadelphia. "Vaccine Ingredients - Aluminum." 4 Nov. 2014, Last Reviewed March 22, 2018. www.chop.edu/centers-programs/vaccine-education-center/vaccine-ingredients/aluminum.

[15]FDA "CFR - Code of Federal Regulations Title 21." Volume 4. Chapter 1, Subchapter C, Part 201, Subpart G, Section 201.323.

[16]O'Shea, *Vaccination is Not,* p. 68.

[17]James, *Immunization: The Reality,* p. 20.

[18]Trebing, Good-Bye Germ, p. 17.

[19]Coulter, *Vaccination, Social Violence,* pp. 156, 157.

[20]Ibid., p. 121.

[21]Mawson, "Preterm Birth, Vaccination and Neurodevelopmental..."

[22]Sproston N, Ashworth J. "Role of C-Reactive Protein at Sites of Inflammation and Infection." *Front Immunol*. 2018. Accessed June 29, 2020. https://www.ncbi.nlm.nih.gov/pmc/articles/PMC5908901/.

[23]Pourcyrous M, et al. Primary Immunization of Premature Infants With Gestational Age <35 Weeks: Cardiorespiratory Complications and C-reactive Protein Responses Associated With Administration of Single and Multiple Separate Vaccines Simultaneously. J Pediatr. 2007. Accessed June 29 2020. https://pubmed.ncbi.nlm.nih.gov/17643770/.

[24]Coulter, *A Shot in,* p. 12.

[25]Ibid., p. 13.

[26]Stewart, G T. "Vaccination against Whooping-Cough. Efficacy..."

[27]James, *Immunization: the Reality,* p. 150.

[28]O'Shea, *Vaccination is Not,* p. 108.

[29]Trebing, *Good-Bye Germ,* p. 74.

[30]Ibid., p. 21.

[31]Ibid., p. 118.

[32]Coulter, *A Shot in,* p. 96.

[33]Ibid.

[34]Coulter, *A Shot in*, pp. 99-100, 101.

[35]Centers for Disease Control and Prevention, "Measles Data and Statistics." Last Updated April 16, 2019, www.cdc.gov/measles/downloads/MeaslesDataAndStatsSlideSet.pdf?s_cid=cs_3520.

[36]Poland, "Failure to Reach the Goal…"

[37]Maready, *Crooked: Man-Made*, pp. 213-214.

[38]Ibid., pp. 204-205.

[39]Ibid., pp. 220-221.

[40]Matthew, et al. "Aluminium in Brain Tissue in Autism." *Journal of Trace Elements in Medicine and Biology*, Urban & Fischer, 26 Nov. 2017, www.sciencedirect.com/science/article/pii/S0946672X17308763.

[41]Children's Medical Safety Research Institute, "Discovery of 'Shockingly High' Levels of Aluminum in Brains of Individuals with Autism Suggests Link with Aluminum-Containing Vaccines," https://www.globenewswire.com/news-release/2017/11/28/1315190/0/en/Discovery-of-Shockingly-High-Levels-of-Aluminum-in-Brains-of-Individuals-with-Autism-Suggests-Link-with-Aluminum-Containing-Vaccines.html.

[42]Maready, *Crooked: Man-Made*, p. 221.

[43]Inbar, "Behavioral Abnormalities in Female Mice..."

[44]Habakus, *Vaccine Epidemic*, p. 288.

[45]Levy, *Curing the Incurable*, pp. 123-124.

[46]Barry, *Vaccine Whistleblower*, p. 109.

[47]Scheibner, *Vaccination: 100 Years*, p. 220.

[48]Ibid., p. 257.

[49]Ibid., pp. 200, 201.

[50]Nguyen, Wilder. *Vaccination a Medical*, pp. 148-149.

[51]O'Shea, *Vaccination is Not*, p. iii.

[52]Maready, *Unvaccinated: Why Growing*, pp. 45-46.

[53]O'Shea, *Vaccination is Not*, p. 59.

[54]Ibid., p. 120.

[55]Ibid., p. 202.

[56]Hume, E. Douglas. *Bechamp or Pasteur: A Lost Chapter in the History of Biology*. 2nd ed., Kessinger Legacy Reprints, 2010. p. 276.

[57]Coulter, *A Shot in*, p. 70.

[58]Scheibner, *Vaccination is Not*, p. 262.

[59]O'Shea, *Vaccination is Not*, pp. 47-48.

[60]Gunn, Vaccination: Its Fallacies, p. 17.

[61]Habakus, *Vaccine Epidemic*, p. 173.

[62]Wilder, *The Fallacy of*, p. 18.

[63]Trebing, *Good-Bye Germ*, p. 192.

[64]Ibid., p. 241.

[65]Ibid., p. 239.

[66]Habakus, *Vaccine Epidemic*, p. 135.

Chapter 7 Seek to the Lord

[1]James, *Immunization: The Reality*, p. 62.

[2]Ibid., p. 57.

[3]Ibid., p. 124.

[4]Ibid., p. 58.

[5]Ibid., p. 104.

[6]Ibid., p. 268.

[7]Ibid., p. 127.

[8]Gregor, *How Not to*, p. 18-19.

[9]Children of God for Life, "U.S. Aborted Fetal Products," https://cogforlife.org/wp-content/uploads/fetalproductsall.pdf

[10]"PER C6 & HEK-293." Children of God for Life, 21 Oct. 2019, cogforlife.org/per-c6-hek-293/#more-776.

[11]Zhang, Feng, et al. "Molecular Mechanism of the Sweet Taste Enhancers." *PNAS*, National Academy of Sciences, 9 Mar. 2010, www.pnas.org/content/107/10/4752.

[12]Maready, *Crooked: Man-Made*, p. 127.

[13]James, *Immunization: the Reality*, p. 210.

[14]Ibid., p. 139.

[15]Ibid., p. 133.

[16]Ibid., p. 134.

[17]Scheibner, 100 Years, p. xi.

[18]James, *Immunization: The Reality*, p. 190.

[19]Wallach, Joel D., and Ma Lan. *Rare Earths: Forbidden Cures*. Double Happiness Pub., 1996. p. 215.

[20]Wallach, Joel D., et al. *Epigenetics the Death of the Genetic Theory of Disease Transmission*. SelectBooks, Inc, 2014. p. 3.

[21]Institute of Medicine. "To Err is Human: Building a Safer Health System." 2000. https://www.ncbi.nlm.nih.gov/pubmed/25077248.

[22]Saul N Weingart, Ross McL Wilson, Robert W Gibberd, and Bernadette Harrison. "Epidemiology of medical error." *BMJ*. 2000. https://www.ncbi.nlm.nih.gov/pmc/articles/PMC1070928/.

"No study rivals the scope of the Harvard and the Australian studies except for a recent replication of the Harvard study in Colorado and Utah. Even so, the results probably represent an estimate of the lower boundary of the prevalence of medical injury and error. The Harvard investigators defined adverse events stringently, using disability and injury as prerequisites. This underestimates the error rate because many errors do not produce injury; they are caught in time, the patient is resilient, or luck is good."

[23]Lazarou , Pomeranz BH, Corey PN. "Incidence of adverse drug reactions in hospitalized patients: a meta-analysis of prospective studies." JAMA . 1998;279(15):1200-5. https://pubmed.ncbi.nlm.nih.gov/9555760/.

"We estimated that in 1994 overall 2216000 (1721000-2711000) hospitalized patients had serious ADRs (Adverse Drug Reaction) and 106000 (76000-137000) had fatal ADRs, making these reactions between the fourth and sixth leading cause of death."

[24]James, *Immunization: The Reality*, p. 268.

[25]Habakus, *Vaccine Epidemic*, p. 274.

[26]Ibid., pp. 127, 128.

[27]Trebing, *Good-Bye Germ*, pp. 188, 189.

[28]Tilden, John H., and Russell T. Trall. *Toxemia Explained/ The True Healing Art*. Life Science Institute, 1989. pp. vi, 35.

[29]Maready, *Crooked: Man-Made*, p. 348.

[30]Tilden, *Toxemia Explained*, pp. 11-12.

[31]Trebing, *Good-Bye Germ*, p. 93.

[32]Levy, *Curing the Incurable*, p. 43.

[33]Ibid.

[34]Humphries, *Dissolving Illusions*, pp. 425-426.

[35]Roschek B; Fink RC; McMichael MD; Li D; Alberte RS; "Elderberry Flavonoids Bind to and Prevent H1N1 Infection in Vitro." *Phytochemistry*, U.S. National Library of Medicine, 2009, pubmed.ncbi.nlm.nih.gov/19682714/.

[36]Zakay-Rones Z; Thom E; Wollan T; Wadstein J. "Randomized Study of the Efficacy and Safety of Oral Elderberry Extract in the Treatment of Influenza A and B Virus Infections." *The Journal of International Medical Research*, U.S. National Library of Medicine, 2004, pubmed.ncbi.nlm.nih.gov/15080016/.

[37]Humphries, *Dissolving Illusions*, p. 428.

[38]Ibid., p. 429.

[39]Ibid., p. 435.

[40]Guyer, "Annual Summary of Vital Statistics..." p. 1315.

[41]Dimova S; Hoet PH; Dinsdale D; Nemery B. "Acetaminophen Decreases Intracellular Glutathione Levels and Modulates Cytokine Production in Human Alveolar Macrophages and Type II Pneumocytes in Vitro." *The International Journal of Biochemistry & Cell Biology,* U.S. National Library of Medicine, 2005, pubmed.ncbi.nlm.nih.gov/15878691/.

[42]Biswas SK; Rahman I. "Environmental Toxicity, Redox Signaling and Lung Inflammation: The Role of Glutathione." *Molecular Aspects of Medicine,* U.S. National Library of Medicine, 2008, pubmed.ncbi.nlm.nih.gov/18760298/.

[43]Schultz ST; Klonoff-Cohen HS; Wingard DL; Akshoomoff NA; Macera CA; Ji M; "Acetaminophen (Paracetamol) Use, Measles-Mumps-Rubella Vaccination, and Autistic Disorder: The Results of a Parent Survey." *Autism : the International Journal of Research and Practice,* U.S. National Library of Medicine, 2008, pubmed.ncbi.nlm.nih.gov/18445737/.

[44]Phillips, Vaccination Myths, pp. 10-11.

[45]Liao, John B. "Viruses and Human Cancer." *The Yale Journal of Biology and Medicine,* Yale Journal of Biology and Medicine, Dec. 2006, www.ncbi.nlm.nih.gov/pmc/articles/PMC1994798/.

[46]Trebing, *Good-Bye Germ,* pp. 80, 81.

[47]Soldatenkova, "An Investigation of Infant Deaths..." p. 1417.

[48]Ibid.

[49]Levy, *Curing the Incurable,* p. 127.

[50]Dimova, "Acetaminophen Decreases Intracellular Glutathione Levels..."

[51]Kalokerinos, *Every Second Child,* pp. 63, 64.

[52]Coulter, *Vaccination, Social Violence,* p. 121.

[53]Coulter, *A Shot in,* p. 85.

[54]Coulter, *Vaccination, Social Violence,* p. 152.

[55]Coulter, *A Shot in,* p. 128.

[56]Ibid., p. 129.

[57]Ibid., p. 125.

[58]Kalokerinos, *Every Second Child,* p. 60.

[59]Coulter, *A Shot in,* p. 118.

[60]Habakus, *Vaccine Epidemic,* p. 308.

[61]Ibid., p. 285.

[62]Kalokerinos, *Every Second Child,* pp. 122-123.

[63]Mullins, *Murder by Injection,* p. 144.

[64]Habakus, *Vaccine Epidemic,* p. 286.

[65]Gunn, Vaccination: Its Fallacies, p. 10.

[66]Dean, Carolyn, and Trueman Tuck. *Death by Modern Medicine*. Matrix Verité, 2005. p. 13.

[67]Parker, David, and Lester, Dawn. *What Really Makes You Ill? Why Everything You Thought You Knew About Disease Is Wrong*. Independently Published, 2019. p. 9.

[68]Mullins, *Murder by Injection*, p. 130.

Chapter 8 Carry Tales to Shed Blood

[1]Kim, Sam S, et al. "Effects of Maternal and Provider Characteristics on up-to-Date Immunization Status of Children Aged 19 to 35 Months." *American Journal of Public Health*, 2007, Feb. 2007, www.ncbi.nlm.nih.gov/pubmed/17194865.

[2]Institute of Medicine (US) Committee to Review the Adverse Consequences of Pertussis and Rubella Vaccines. "Evidence Concerning Pertussis Vaccines and…"

[3]Dehlendorf, Christine, et al. "Disparities in Abortion Rates: a Public Health Approach." *American Journal of Public Health*, American Public Health Association, Oct. 2013, www.ncbi.nlm.nih.gov/pmc/articles/PMC3780732/.

[4]Hak, "Negative Attitude of Highly Educated…"

[5]Gullion, Jessica Smartt, et al. "Deciding to Opt out of Childhood Vaccination Mandates." *Public Health Nursing (Boston, Mass.)*, U.S. National Library of Medicine, 2008, www.ncbi.nlm.nih.gov/pubmed/18816357.

[6]"A Quote by Malcolm X." Goodreads, www.goodreads.com/quotes/8869214-the-white-liberal-is-the-worst-enemy-to-america-and.

[7]Nguyen, Wilder, *Vaccination a Medical*, p. 143.

[8]Ibid., p. 155.

Chapter 9 Choose You This Day Whom Ye Will Serve

[1]Ewing, "What Is Regressive Autism and...?"

[2]O'Shea, *Vaccination is Not*, p. 107.

Chapter 10 Nehushtan

[1]Emery, Kate Meyers. "New Morbid Terminology: Corpse Medicine." Bones Don't Lie, 17 Apr. 2015, bonesdontlie.wordpress.com/2015/04/08/new-morbid-terminology-corpse-medicine/.

[2]Davies, Owen, and Francesca Matteoni. "'A Virtue beyond All Medicine': The Hanged Man's Hand, Gallows Tradition and Healing in Eighteenth- and Nineteenth-Century England." *Social History of Medicine : the Journal of the Society for the Social History of Medicine*, Oxford University Press, Nov. 2015, www.ncbi.nlm.nih.gov/pmc/articles/PMC4623855/.

[3]Bieler, Henry G. *Food Is Your Best Medicine*. Ballantine Books, 1992. pp. 32-33.

[4]Davies et al, "'A Virtue beyond All Medicine...'"

[5]Dio, Cassius, and E. Cary. *Dio's Roman History IX Books LXXI-LXXX*. Vol. 9, Wm. Heinemann, 1927.

p. 101.

[6]Peebles, Vaccination, A Curse, pp. 87-88.

[7]Ibid., pp. 238-239.

[8]Prakash, M, and J Carlton Johnny. "Things you don't learn in medical school: Caduceus." *Journal of pharmacy & bioallied sciences* vol. 7, Suppl 1 (2015): S49-50. doi:10.4103/0975-7406.155794

[9]Stringfellow, Alan B. *Insights on the Book of Revelation: a Verse by Verse Study*. Whitaker House, 2014. p. 48.

[10]"Forsaking God For the Sake of Science." Children of God for Life, 3 Oct. 2019, cogforlife.org/2012/06/13/polioperversion/.

[11]Mullins, *Murder by Injection*, p. 66.

[12]James, *Immunization: The Reality*, pp. 148-149.

[13]Trebing, *Good-Bye Germ*, pp. 68-69.

[14]Coulter, *A Shot in*, p. 193.

[15]Jones, Floyd Nolen. *Which Version Is the Bible?* 21st ed., KingsWord Press, 2018. pp. 199-200.

[16]Tilden, *Toxemia Explained*, p. 21.

Chapter 11 Days of Noah

[1]O'Shea, *Vaccination Is Not*, p. 28.

[2]Maready, *Moth in the*, p. 156.

[3]Coulter, *Vaccination, Social Violence*, p. 157.

[4]Hume, *Bechamp or Pasteur*, p. 276.

[5]Bovine Derived Materials Used in Vaccine Manufacturing Questions and Answers. U.S. Food and Drug Administration. https://www.fda.gov/vaccines-blood-biologics/questions-about-vaccines/bovine-derived-materials-used-vaccine-manufacturing-questions-and-answers.

[6]Institute of Medicine (US) Immunization Safety Review Committee." Immunization Safety Review: SV40 Contamination of Polio Vaccine and Cancer". Edited by Kathleen Stratton et al., *National Academies Press,* 2002, www.ncbi.nlm.nih.gov/pubmed/25057632.

[7]Vilchez, Regis A, and Janet S Butel. "Emergent Human Pathogen Simian Virus 40 and Its Role in Cancer." *Clinical Microbiology Reviews,* American Society for Microbiology, July 2004, www.ncbi.nlm.nih.gov/pmc/articles/PMC452549/.

[8]Cashman, N R. "Transmissible Spongiform Encephalopathies: Vaccine Issues." *Developments in Biologicals,* U.S. National Library of Medicine, 2001, www.ncbi.nlm.nih.gov/pubmed/11761262.

[9]Piccardo P, Cervenakova L, Vasilyeva I, et al. "Candidate Cell Substrates, Vaccine Production, and Transmissible Spongiform Encephalopathies." *Emerging Infectious Diseases.* 2011;17(12):2262-2269. doi:10.3201/eid1712.110607. https://www.ncbi.nlm.nih.gov/pmc/articles/PMC3311205/.

[10]Phillips, Vaccination Myths, p. 15.

[11]O'Shea, *Vaccination is Not,* pp. 56, 58.

[12]James, *Immunization: The Reality,* p. 83.

[13]Walene, *Immunization: The Reality,* p. 20.

[14]Coulter, *Vaccination, Social Violence,* p. 157.

[15]Coulter, A Shot in, p. 150.

[16]Coulter, *Vaccination, Social Violence,* p. 144.

[17]Ibid., p. 181.

[18]Ibid., p. 198.

[19]Ibid., pp. 129-130.

[20]Coulter, A Shot in, p. 219.

[21]Coulter, *Vaccination, Social Violence,* p. 45.

[22]Coulter, *Vaccination, Social Violence,* pp. 36-37.

[23]Ibid., p. xiv.

[24]Trebing, *Good-Bye Germ,* p. 63.

[25]Coulter, *A Shot in,* p. 208.

[26]Ibid., p. 206.

[27]Coulter, *Vaccination, Social Violence,* p. 94.

[28]Jacqueline. "Science: Fetal DNA Cross-Reacting W/ Kids' DNA & Plotkin's Words (Video)." *Deep Roots at Home,* 3 Sept. 2021, https://deeprootsathome.com/science-dna-cross-reacts-with-kids-dna/.

[29]Jacqueline. "Science: Fetal DNA Cross-Reacting W/ Kids' DNA & Plotkin's Words (Video)." *Deep Roots at Home,* 3 Sept. 2021, https://deeprootsathome.com/science-dna-cross-reacts-with-kids-dna/.

[30]"Vaccines Did Not Save Humanity and Never Will – Dr Suzanne Humphreys MD." *Vaccination Information Network,* 30 Apr. 2016, www.vaccinationinformationnetwork.com/vaccines-did-not-save-humanity-and-never-will-dr-suzanne-humphreys-md/.

[31]Gates, Bill. 'Innovating to Zero!" *TED,* 2010, www.ted.com/talks/bill_gates_innovating_to_zero.
https://www.ted.com/talks/bill_gates_innovating_to_zero/discussion.

To go right to the transcript visit: https://www.ted.com/talks/bill_gates_innovating_to_zero/transcript#t-278552.

[32]"Bill Gates on the anti-vaccine movement" from CBC News: The National, Bill Gates speaks to Peter Mansbridge about the anti-vaccination movement. https://www.youtube.com/watch?v=dZSL8OnBLlc.

[33]O'Shea, *Vaccination is Not*, p. 55.

[34]Ibid., pp. 200-201.

[35]Ibid., p. 198.

[36]Horowitz, Leonard G. *Emerging Viruses: AIDS and Ebola: Nature, Accident, or Intentional?* Tetrahedron Press, 2014. p. v.

[37]Peebles, *Vaccination, A Curse*, p. 117.

[38]Mardis, Jeffrey W. *What Dwells beyond: the Bible Believer's Handbook to Understanding Life in the Universe*. 3rd ed., Defender Publishing, 2015. pp. 99, 467.

[39]Project Veritas. "Pfizer Whistleblower Leaks Execs Emails: 'We Want to Avoid Having Info on Fetal Cells out There'." *YouTube*, YouTube, 6 Oct. 2021, https://www.youtube.com/watch?v=FUXGB5FzhPc.

[40]North Dakota Health. Health.nd.gov. "COVID-19 Vaccines & Fetal Cell Lines." https://www.health.nd.gov/sites/www/files/documents/COVID%20Vaccine%20Page/COVID-19%20Vaccine%20Fetal%20Cell%20Handout.pdf

[41]Children of God for Life. "Guidance on Getting the COVID-19 Vaccine - Children of God for Life." *Children of God for Life - Pro-Life Leader in Campaign for Ethical Research*, 6 Oct. 2021, https://cogforlife.org/guidance/.

[42]Jacqueline. "CDC Data: 16,310 Deaths/778,683 Other Injuries, Adv Reactions Reported." Deep Roots at Home, 11 Oct. 2021, https://deeprootsathome.com/early-cdc-warning-list-of-severe-adverse-reactions/.

[43]Maiti S, Banerjee A, Kanwar M. In silico Nigellidine (N. sativa) bind to viral spike/active-sites of ACE1/2, AT1/2 to prevent COVID-19 induced vaso-tumult/vascular-damage/comorbidity. Vascul Pharmacol. 2021 Jun;138:106856. doi: 10.1016/j.vph.2021.106856. Epub 2021 Mar 18. PMID: 33746069; PMCID: PMC7970800.

[44]Jacqueline. "Black Seed 'Strongly Binds' to Spike Protein, Prevents Vascular Damage." Deep Roots at Home, 29 Sept. 2021, https://deeprootsathome.com/black-seed-binds-to-spike-protein-prevents-vascular-damage/.

[45]Common dandelion (Taraxacum officinale) efficiently blocks the interaction between ACE2 cell surface receptor and SARS-CoV-2 spike protein D614, mutants D614G, N501Y, K417N and E484K in vitro. Hoai Thi Thu Tran, Nguyen Phan Khoi Le, Michael Gigl, Corinna Dawid, Evelyn Lamy. bioRxiv 2021.03.19.435959; doi: https://doi.org/10.1101/2021.03.19.435959

[46]Jacqueline. "Dandelion Tea May Block Spike Protein on Syncitin 1, Prevent Miscarriage." Deep Roots at Home, 24 Sept. 2021, https://deeprootsathome.com/can-this-1-herb-prevent-miscarriage/.

[47]Bruno, Roxana. "Why Covid-19 Vaccines Might Affect Fertility." CienciaySaludNatural.com, https://cienciaysaludnatural.com/estudio/why-covid-19-vaccines-might-affect-fertility/.

[48]Jacqueline. "Dandelion Tea May Block Spike Protein on Syncitin 1, Prevent Miscarriage." Deep Roots at Home, 24 Sept. 2021, https://deeprootsathome.com/can-this-1-herb-prevent-miscarriage/.

[49]Bernadetta Lis, Agata Rolnik, Dariusz Jedrejek, Agata Soluch, Anna Stochmal, Beata Olas, Dandelion (Taraxacum officinale L.) root components exhibit anti-oxidative and antiplatelet action in an in vitro study, Journal of Functional Foods, Volume 59, 2019, Pages 16-24, ISSN 1756-4646, https://doi.org/10.1016/j.jff.2019.05.019.

[50] Huff, Ethan, and Sherri Häusler. "Miscarriages and Cancer up 300%, Neurological Problems up 1,000% Due to Covid 'Vaccines.'" *The Truth About Vaccines*, 2 Feb. 2022, https://thetruthaboutvaccines.com/covid-vaccine-miscarriages-cancer/.

[51]"Regeneron Position Statement on Stem Cell Research." Regeneron.com, Apr. 2020, https://www.regeneron.com/downloads/regeneron-position-stem-cell-research.pdf.

[52]Trebing, *Good-Bye Germ*, p. 69.

Chapter 12 Speak the Truth in Love

[1]Gunn, *Vaccination: Its Fallacies*, p. 32.

[2]Ibid., p. 27.

[3]Ibid., p. 36.

[4]Althen v. Sec'y of HHS, 418 F.3d 1274.

[5]James, *Immunization: the Reality*, p. 106.

[6]McBean, The Poisoned Needle, p. 77.

[7]Huff, *How to Lie*, p. 100.

[8]O'Shea, *Vaccination Is Not*, p 17.

[9]Greger, *How Not to*, p. 1.

[10]Gunn, *Vaccination: Its Fallacies*, p. 31.

[11]Ibid., p. 3.

[12]Peebles, *Vaccination: a Curse*, p. 299.

[13]Hodge, *The Vaccination Superstition*, pp. 26-27, 43-44.

[14]McBean, *The Poisoned Needle*, p. 35.

[15]Ibid., p. 158.

[16]Nguyen, *Vaccines: the Biggest*, Loc 79.

[17]McBean, *The Poisoned Needle*, p. 105.

[18]Nguyen, *Vaccines: the Biggest*, Loc 2.

[19]Mendelsohn, *How to Raise*, p. 230.

[20]Humphries, *Dissolving Illusions*, p. 478.

[21]Gunn, *Vaccination: Its Fallacies*, p. 21.

[22]Ibid., p. 37.

[23]O'Shea, *Vaccination Is Not*, p. 188.

[24]Ibid., p. 200.

[25]Wolfe, "Anti-Vaccinationists Past and Present."

[26]James, Immunization: The Reality, p. 243.

[27]Peebles, Vaccination, A Curse, pp. 5,6.

[28]Gunn, *Vaccination: Its Fallacies*, pp. 30, 31.

[29]Wolfe, "Anti-Vaccinationists Past and Present."

[30]Nguyen, Wilder, *Vaccination a Medical*, p. 191.

Chapter 13 Give Place

[1]Coulter, *A Shot in*, p. 134.

[2]Ibid., p. 138.

[3]Ibid., pp. 179-180.

[4]Ibid., p. 17.

[5]Ibid., p. 19.

[6]Ibid., p. 21.

[7]Ibid., pp. 219-220.

[8]Ibid., p. 217.

[9]Ibid., p. 27.

[10]Ibid., p. 63.

Chapter 14 In Conclusion—Mercy and Truth

[1]"6 Amazing Scientific Facts of the Bible | Proof for God." World Video Bible School. https://www.youtube.com/watch?v=EvOWr83qYk0

[2]"5 Prophecies that Prove the Bible | Proof for God." World Video Bible School. https://www.youtube.com/watch?v=PgN33jH6jHk&feature=youtu.be

[3]"9 Discoveries that Confirm the Bible | Proof for God." World Video Bible School. https://www.youtube.com/watch?v=oQd-fdpq9mk&t=9s

[4]"7 Superhuman Qualities of the Bible | Proof for God." World Video Bible School. https://www.youtube.com/watch?v=yWLhVkOFcYk

[5]"What about Contradictions in the Bible? | Proof for God." World Video Bible School (WVBS). https://www.youtube.com/watch?v=NkBbTIFBl5I&t=5s

[6]Stoner, Peter Winebrenner., and Robert C. Newman. *Science Speaks: Scientific Proof of the Accuracy of Prophecy and the Bible.* Moody Press, 1976.

[7]Gray, Jonathan. *Dead Men's Secrets: Tantalising Hints of a Lost Super Race*. Vision Australia Information Library Service, 2015. pp. 52, 53.

[8]Jones, *Which Version Is,* pp. 175, 177.

[9]Gray, *Dead Men's Secrets,* pp. 50, 51.

[10]Dead Men's Secrets, p. 70

[11]Ibid. p. 16.

Made in the USA
Monee, IL
04 November 2026

67382832R00174